Trends Waves Windows & Bubbles

Kenneth J. Thurber Ph.D.

digital
systems
press

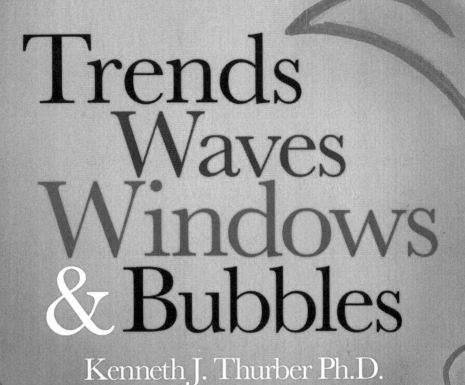

Trends Waves Windows & Bubbles

Kenneth J. Thurber Ph.D.

TRENDS, WAVES, WINDOWS & BUBBLES
Copyright © 2014 by Kenneth J. Thurber.
All rights reserved.

ISBN 13: 978-0-9833424-6-5

Library of Congress Control Number: 2014931326

Edited by Paul O'Neill
Cover Illustration by Shawn McCann
Book design and typesetting by James Monroe Design, LLC.

Digital Systems Press

33 Thornwood Drive, Suite 500
Ithaca, NY 14850
–and–
9971 Valley View Road
Eden Prairie, MN 55344

www.digitalsystemspress.com

To my wife.

CONTENTS

Acknowledgments . . . . *vii*
List of Figures . . . . *ix*
Preface . . . . *xi*
Prologue . . . . *xiii*

PART 1: TRENDS 1

Chapter 1: *Trend Concepts* *3*
Chapter 2: *Trends and their Importance* *7*
Chapter 3: *Trend Types* *11*
Chapter 4: *Definition of Trends* *13*
Chapter 5: *Visualization of Trends* *17*
Chapter 6: *Problems with Trends* *21*
Chapter 7: *Decay of Trends* *25*
Chapter 8: *CPT—Death of a Trend* *29*
Chapter 9: *Variations on a Trend* *33*
Chapter 10: *Reinvention of Trends* *37*

PART 2: WAVES 41

Chapter 11: *Wave Concepts* *45*
Chapter 12: *Importance of Waves* *49*
Chapter 13: *Wave Types* *53*
Chapter 14: *Wave Definitions* *57*
Chapter 15: *Visualization of Waves* *61*
Chapter 16: *Problems with Waves* *65*
Chapter 17: *Decay of Waves* *69*

Chapter 18: *Death of a Wave* 71
Chapter 19: *Variations of a Wave* 75
Chapter 20: *Reinvention of Waves* 77

PART 3: WINDOWS 81

Chapter 21: *The Concept of Windows* 85
Chapter 22: *The Importance of Timing the Window* 89
Chapter 23: *Window Types* 91
Chapter 24: *Definition of Windows* 95
Chapter 25: *Visualization of Windows* 99
Chapter 26: *Problems with Windows* 103
Chapter 27: *Decay of Windows* 107
Chapter 28: *Death of a Window* 111
Chapter 29: *Window Variations* 115
Chapter 30: *Reinvention of Windows* 117

PART 4: BUBBLES 121

Chapter 31: *Concept of Bubbles* 125
Chapter 32: *Importance of Bubbles* 127
Chapter 33: *Types of Bubbles* 131
Chapter 34: *Definition of Bubbles* 133
Chapter 35: *Visualization of Bubbles* 135
Chapter 36: *Problems with Bubbles* 139
Chapter 37: *Decay of Bubbles* 143
Chapter 38: *Death of a Bubble* 147
Chapter 39: *Variations of a Bubble* 151
Chapter 40: *Reinvention of Bubbles* 155

Epilogue *159*
Index *163*

ACKNOWLEDGMENTS

A number of people contributed to the development and publication of this book. Valuable contributions to the concepts in this book were made by Barry Trent, Julie Baker, Gene Proctor, John Metil and Noel Schmidt. The actual production of the book was ably completed by Shawn McCann and James Monroe. Editing and coordination of the project, as well as, insightful comments and critiques of the book were provided by Paul O'Neill. Julie Baker produced both the softback version and the e-version of the book.

LIST OF FIGURES

Figure 1: Illustration of a simple semi-log graph with a hockey stick pattern product launch

Figure 2: Illustration of the growth of Transistor Density on Single Chips over Time – Moore's Law

Figure 3: Mainframe Centric Data Processing Center circa 1965

Figure 4: CPT 8000 Series

Figure 5: Chairman Mao

Figure 6: Illustration of the growth of internet hosts from 1981 until 1998.

Figure 7: Illustration of the growth of internet search.

Figure 8: Illustration of the growth of FB users of social media

Figure 9: Illustration of Housing starts in the US from 2003 through 2012

PREFACE

Trends, Waves, Windows & Bubbles is a book about technology big waves. It's the third in a series (the first two were *Big Wave Surfing* and *Do Not Invent Buggy Whips*) looking at how technology moves from the research lab and into viable products. How do ideas become inventions? How do inventions become products? More importantly, how can you become involved in that process?

My first book, *Big Wave Surfing*, talked about finding and managing high-growth companies.

Do Not Invent Buggy Whips looked at the creation of disruptive ideas so that you can create the wave that is spotted by others.

In *Trends, Waves, Windows & Bubbles*, we're going to look at how to spot, not necessarily create, a wave. For a variety of reasons, not everyone can create a disruptive technology wave. However, there is no reason that you cannot profit from waves created by others. If you can spot a wave, you can profit.

We'll take a top-down approach to help you spot a wave.

First, we'll examine the idea of a trend. By spotting trends you have a chance to identify opportunities and then try to capitalize on them.

Once you spot a trend, the next step is to determine if it can cause a wave of product innovation. If a trend causes a wave then you have a chance to spot viable opportunities for either new products or for making money on emerging technology companies.

If you spot a wave you need to decide if that has created an opportunity or window that you can enter. This brings us to the window of opportunity. The primary issue is how you can spot an actionable window of opportunity.

Lastly, we do not want to get crushed or participate in a false start. So, we want to look at what I would call a bubble. Bubbles exist because once a wave and window of opportunity form there is a tendency for money to rapidly flow into the space and the entire opportunity, if viable, can quickly evaporate.

Now you understand the title of the book: *Trends, Waves, Windows & Bubbles*.

The primary goal here is to examine the issues associated with the process of spotting (not creating) a big wave that will allow us to profit from our insight and analysis. We are looking at this as a way to make decisions. Some of the decisions may be investment decisions. Some of the decisions may involve where we might find interesting work. But, our focus is on spotting, not creating technology waves.

For symmetry this book is divided into four parts: one part for each issue. Each part has an introduction, specific chapters and a conclusion. There's a prologue and an epilogue. It's a modular structure. For people with busy schedules you can read the book in small bites. Others will enjoy in one sitting. As Julia Child would say "Bon Appetit."

PROLOGUE

After writing my first business book, *Big Wave Surfing*, I was motivated by people's questions to elaborate on some of the issues discussed in the book. The result was the book *Do Not Invent Buggy Whips*. This book, *Trends, Waves, Windows & Bubbles* is a further look at the process of spotting and riding big waves.

Whereas, *Big Wave Surfing* provided an overview of the subject of spotting and riding big technology waves, here we'll try to spot big waves formed by trends. We'll examine techniques that will help us determine if a trend is significant and if it will create a big wave.

The book, *Do Not Invent Buggy Whips*, led the reader through the process of creating (or trying to create) a big wave. Not everyone is capable of creating a big wave, so in this book we'll concentrate on how to spot a big wave that has been created, not on creating your own wave! However, even if a trend leads to a big wave you must catch the wave (when a window of opportunity opens up) and then you must get off the wave (before a bubble occurs). Timing is critical. Timing can easily mean the difference between success (profits) and disaster! Thus, the topics discussed in this book are: trends, waves, windows and bubbles.

The purpose of *Big Wave Surfing* was to provide an overview of the concepts of technology big wave surfing. The first part of the book focused on spotting big waves. So we started with the idea of a disruption, proceeded to the issue of trade-offs and finished with the concept of how big waves form. In the second half of the book we looked at how you could ride a big technology wave into product development, management, marketing and investing. The idea was to provide an end-to-end view of the process as well as the risks and rewards of technology big wave surfing.

The purpose of *Do Not Invent Buggy Whips* was to look at one particular aspect of big wave surfing; i.e., how do you go about the process of creating a big wave. The focus was on creating innovative ideas and positioning them in the marketplace. One key to big wave surfing is spotting the disruption that causes a big wave to form. Real surfers are aided in this process by weather and ocean swell reports available from a variety of sources. So they have the opportunity to spot big waves before they form and move onto land. Technology big wave surfers can do the same using trend analysis (discussed in this book) but they can also try to create their own big waves. Although creation of a big wave was the subject of *Do Not Invent Buggy Whips*, not everyone will be able to grasp the concepts that will create a big wave. However, you do not have to be able to create a wave to profit from its existence.

Even though you do not have to create a big wave to be successful and profit from the technology big wave phenomena, you do need to be able to recognize the possibility of a big wave and determine when or if it becomes a big wave. At that point, you need to figure out what to do about your discovery. Unlike the previous books which were not explicitly about making money, here we do want to identify and exploit trends and waves by identifying a window of opportunity. The key concept to spotting a big wave is spotting trends. If you are a surfer this is when you catch and ride your wave. If you are in the investment or product business spotting your window of opportunity is when you can make money. You should think of this book as focused on finding the window of opportunity and then hopping on the resultant technology to capitalize it—and make money! There is no veneer of sophistication in the premise of this book. The reason to spot trends and waves is to find the window of opportunity and make money. At the same time we want to avoid getting hammered when a technology bubble bursts.

Spotting a trend is a very difficult task. You can never be sure that it is real until it is far into its potential life cycle. You may find it difficult to conceptualize a big wave if it's a change from the norm. It seems to be human nature to be reticent to make substantial changes to lifestyles and habits and this will effectively slow down recognition of a big wave if it involves substantial change. In the case of defining substantial change we are talking about any change that's beyond your view of the norm.

But, spotting a trend is only one part of the equation. The other part is what you are going to do about it. In some cases you do not care that a trend exists. If the trend is for children to only play on grassy surfaces, there is no impact to your life when your children play on your large, grassy lawn. On the other hand if the trend is for gas prices to increase rapidly and you commute fifty miles to work every day in a gas guzzler, you might think about making some changes to your lifestyle. Importantly, you can only make choices when you recognize and act on a trend. If you just bought a large SUV and the next day gas prices shoot up, you are not in a good position to make choices about how and when to get an economical car. Yet, the general trend is for gas prices to go up and you should not be surprised about an increase in the price of gas. You may have misjudged the rate of change and a sudden spike in price (unless it reverses) would be difficult to deal with over an extended period of time.

It is important to recognize that some trend changes are not important, but some are vital. If you are close to retirement and the rules for taxes on retirement plans change dramatically, you have a problem. If you are just starting your work career then you should have plenty of time to make changes. Retirement planning is a very difficult problem. If you look at the current trends in retirement strategies, you need to ask yourself how you can make a long term plan for retirement. The current tax structure encourages tax deferred savings for retirement, but the law is changing every year. You have no guarantees that your current planning will still work thirty years from now. The importance of knowing when a trend or wave is changing and when to abandon your approach and

TRENDS, WAVES, WINDOWS & BUBBLES

reposition your investments is critical. Timing changes between phases (trends to waves to windows to bubbles) is critical to success.

You must not only understand a trend but the underlying assumptions that are causing the trend.

What works today probably does not work tomorrow. Unless the trend continues!

This book will examine the idea of trends and how they relate to spotting big waves particularly technology waves. It is organized into four parts. There is a discussion of trends, a look at why some waves are more important than others, what to do when you spot a big wave and how waves turn into bubbles.

Those four parts translate into our look at the concepts of trends, waves, windows and bubbles.

Trends and analysis of trends will be our tools to try and find waves that are forming. If we can spot a trend we will look closely to see if a wave can be found. If we can identify a wave, the question becomes can we catch the wave. A window of opportunity may open up and allow us to identify the trend and catch the wave. However, caution is in order because if a really big wave forms, investment money may flow in so fast that a bubble forms. So as we move to the last part of the book we want to take a closer look at the characteristics of bubbles—how they form and how you can avoid their eventual collapse.

Before we get into the meat of the book, I do have to give you my obligatory bias disclaimer. I have been in the electrical and computer engineering business for all of my adult life. When I need an example to discuss in the book, I will usually look to examples that I'm am familiar with from my past. Sometimes I will pick out issues from fields that are far from electronics. But, when it comes down to examples, I usually take the easiest and most comfortable route for me. Since electronics and its derived products are everywhere in society I do not view that strategy as a limitation of people's ability to understand the book.

PART 1
TRENDS

Introduction to Trends

Trends are important phenomena. They exist all around us and in a sense define us. They can appear in many forms.

Look at the spread of a simple virus like the flu. As flu season approaches you can track it as it moves across countries and see the growth trend increase by the number of people infected as time passes. Then, with luck the infection runs its course and everything is back to normal.

Take another example that many of us have encountered. Consider the trend created by the shortage and the problem of getting your hands on the "hot" toy of the Christmas season. In this example the shortage can cause the trend to extend for a longer duration than a simple trend like the flu infection.

These are just two examples of simple trends. The trend that we are most interested in is a trend that has the possibility of creating a big wave. We want to find the trend that has the possibility of creating investment, product or corporate opportunities from which business or sales can occur. For investment and excitement we want to find a big wave—a trend that will have rapid technology growth. When you have rapid technology growth you have the potential to exploit the technology wave for economic gain. This is when you find the window of opportunity and ride the wave before it collapses or the inevitable bubble forms and breaks.

TRENDS, WAVES, WINDOWS & BUBBLES

There are different types of trends. Some are long term, like the trend towards digital electronics over analog electronics. Some are of a relatively short duration, like the trend in specialized Internet applications. Digital electronics is a long term trend with many sub-trends. Internet trends can form rapidly and disappear fast if a dominant vendor shows up and the competitors move on to new endeavors. Even within the long-term growth of digital electronics there have been multiple sub-trends and these trends have varying duration. A classic example of a trend is Moore's Law. Simply stated Moore's Law says that the number of transistors on integrated circuits doubles every two years. A variation of this law says that speed and performance of a chip doubles approximately every 18 months.

Various authors have different views of waves and trends. Sometimes they also use different terminology like windows of opportunity, product opening and disruptive products to describe trends. We will focus our terminology on the concepts of trends and waves. In this book, we will not be looking solely at long term trends, but instead we will look at trends as the fundamental tipping point that causes a big wave. We do not have to be the person involved in wave creation; we just need to be able to spot the wave.

We do not have to be early in the trend but we need to be able to spot a trend because a trend can generate a lot of waves. For example, the invention of the internet has caused a series of waves that continue to this day. You did not have to invent the internet or even spot the initial innovative applications to profit; but once you identify the internet as a trend generating big waves you need to pay attention to determine where and when the next big wave will form.

This part of the book on trends will define and illustrate the basics of trends. The basics we develop here will come in handy later when we determine whether a trend is forming and if the trend can form a big wave that we can exploit.

Chapters 1 to Chapter 10 that follow will elaborate further.

Chapter 1
Trend Concepts

The concept of a trend is very simple. I would call it a set of actions and interactions that define a repetitive pattern or behavior; it is really just a general direction (or tendency) that something is moving in like the current style in fashion.

We see trends around us all the time. Think of a traffic pattern. Traffic builds and cars slow down as rush hour approaches. The increase in traffic is a trend. Interestingly, the resultant slowdown on the freeway is also a trend and the two trends are related. As traffic increases, speed decreases and there is a relationship between the two phenomena. These two trends can be modeled and measured. They are consistent enough that the models can make traffic flow better through use of ramp metering. In some cities, there are actually dynamic traffic models that meter traffic to try to decrease the delay associated with the trend of increasing traffic and the resultant traffic slowdown. We can observe the traffic and consider it similar to disruptive technology in that the patterns of disruption created by traffic are created by the actions of outside events. And like traffic patterns if we want to profit from disruptive technology that others create we need to be alert and work at spotting trends that can spawn a big wave or series of big waves.

Trends exist in many forms. A trend may be as simple as some new fashion worn by teenagers. Or, it may be as complex as a new potential form of computing—quantum computing for example. Quantum computing is a new theory of how to build a computer based upon quantum mechanics to harness the properties of atoms and molecules to perform memory and processing tasks. In many ways it's still an idea before it's time and the researchers are still considering what is really possible. Even the proponents believe that the possibility of success is very far in the future.

If we are looking to spot a trend and use that as the basis of a business decision, the potential of the trend as well as the timing of the trend is of critical importance. We need to figure out the timing of the trend and whether it will actually take off and run for a reasonable length of time. This skill is critical to success. The great investor Peter Lynch of Fidelity investments used to walk around shopping centers as a way to detect and understand new long-term trends in retailing and thus get new stock purchase ideas. In a large number of cases spotting a trend is just that simple. Observe what is happening in the world and ask yourself if the events of the day will cause changes to the behavior or purchasing plans to the masses. A trend that is ongoing (but its results are not clear) is the earlier and earlier opening of stores during the Christmas shopping season. Will this trend cause a disruption in the retail space? What is the long term impact of cyber shopping during the Christmas season? Will Amazon become dominant in the retail space? These are all questions about trends (from simple observations) that one should ask if you are running a retail store or intend to invest in the retail space.

To get started we need a simple definition of a trend. We will define a trend as simply a time-based set of observations which tend to move in a particular direction. For now this simple definition will suffice. When we get to Chapter 3 we will define a trend in detail and examine its component parts. But for now we will use this simple definition.

At this point we only want to establish a basis to examine the properties of a trend. We need to think about trends as a reference that can help us determine anything from the best time to drive on a freeway to the cost of something we might want to purchase. Once we understand the idea of a trend we can then move into the process of figuring out general characteristics of trends and what type of a trend will be of interest to us.

Generally we want to look at a trend in terms of something that we can measure. Unless we can measure a trend we will have difficulty recognizing the trend. So we need some form of measurement to determine the characteristics of a trend.

Chapter 2
Trends and Their Importance

When we are about to engage in a business activity we want to be successful. We do not want to start an enterprise or develop a product that has little chance of success.

If a market is in decline then we do not want to enter that market. Similarly, if there is no market for a product we do not want to enter the market. But, if there is a new product category forming and it looks as if there will be significant possibility of sales, then we want to consider entering that market for that product category.

In many cases, the most successful company or product is not the first product into the space!

I can cite chapter and verse about products (companies) that were the first mover or prime mover but by the time the markets began to move they were in an inferior product position and were quickly aced out by a fast-reacting, nimble, competitor. Take Visi-Calc, (short for Visible Calculator) created in 1979 by Dan Bricklin and Bob Frankston and developed through their company, Software Arts. It was the original spreadsheet developed for the personal computer and many people called it the "software tail" that wagged the personal computer dog. It was a breakout piece of software, an application that gave people a reason to buy personal computers. Not long after, SuperCalc, Lotus 1-2-3, Multiplan from Microsoft, AppleWorks and eventual market leader, Excel from Microsoft, showed up and VisiCalc was history. It had a wild ride for a couple of years. Lotus eventually bought them, the founders got wealthy, and VisiCalc, the original spreadsheet program for the personal computer industry, is now a footnote in history.

There are also cases of people who invented products that were so far ahead of the curve that by the time the market had developed their product or concept was obsolete.

You can never tell how the market will work out in the short term. You also cannot figure out how it will sort itself out long-term. However, the longer a trend runs the more likely the market leaders change or some event creates a trend that allows us the opportunity to position a product for sale.

Consider the stock market for example. You can hear the talking heads pontificating about average movements, but what does that mean unless you own a faithful representation of the average. If you own single stocks then you will see a different trend (or set of trends) than if you own the average. The average is made up by combining a series of trends (or stocks) to form the average. If your portfolio consists of two ten dollar stocks and one is up two dollars and the other is down two dollars, the portfolio has the same value as at the start of the day. To conclude that your portfolio is stable is a simple conclusion based upon its "average change for the day". Yet there was significant change in the individual components making up your portfolio. They were moving within their own trends.

So there are trends within a trend. Some trends can be short-lived and some have a long life expectancy. Some trends will allow for small movements and other trends may make significant movements. If we go back to Moore's Law, we need to ascertain if the law is a general law that applies to all chips or if it is an average. For example, we should try to determine if memory chips conform to the law. Because of their repetitive nature, memory chips probably have an easier time of achieving a greater growth rate than processor chips because they can be built with a large amount of repetitive structure. Yet, if we look at solid state disk drive (SSD) technology we realize that not all of the capacity of the SSD can be applied in an application. We need some amount of redundant memory so that a small failure does not fail the entire device's capabilities. If we are trying to determine how SSD technology fits into a trend, we need to determine whether we measure the transistors or the actual usable memory capacity or whether these two are really the same.

Chapter 3
Trend Types

As we think about trends we need to consider whether there are different types and what properties they could have.

We need to consider the following questions:

1. Are there macro trends and micro trends? If so, what is the difference?

2. Can a trend exist within a larger trend? Can multiple trends exist within a larger trend?

3. Do trends have duration?

4. How do we measure a trend?

5. If trends are made up of components, can we determine the relative significance of the components?

These are just some of the questions that we could ask.

The problem is that trends are a very complex subject. To make our job simpler we will limit our view to a series of parameters that will encompass the major portions of a trend that we can understand. The more detail that we put into the analysis the more we stand to lose the basic structure that will help us to determine whether we have seen or encountered a trend that's relevant to a product or financial decision.

There are many types of trends. But, if we get too caught up in a detailed analysis we probably could do a good job of describing the details but we'd probably land in "analysis by paralysis" territory and find ourselves unable to determine an overall direction.

A classic trend I encountered was the growth rate of IBM during the mid-eighties. For a period of years IBM was consistently growing at a compounded rate of 15%. During that time they may have had some variation in the growth rate from quarter to quarter, but the overall rate for a period of years was 15%. If you did not have some sort of averaging mechanism you could miss the overall trend (15%) and either over-estimate or underestimate the trend. However, the big question was how soon you could spot the trend. And, a very interesting question was whether you could spot the end of the trend. The reason that this trend and its end were so interesting is that big money could be made on Wall Street by spotting the trend. When the trend ended and IBM fell on hard times, the money lost was potentially huge. Another interesting aspect of this trend was trying to determine the root cause of the trend. Was IBM really a superior company or was it just lucky? Were there any underlying technology changes that were facilitating the trend? Depending upon these answers, could a trend spotter find other companies or products that would take advantage of the underlying technology change? One possible answer for the trend was the overall movement to the personal computer (PC) which goosed sales for a period of time before mainframe and minicomputer sales were cannibalized by the rapid adoption of the PC and its lower priced software.

The growth rate of IBM was really a composite trend. It was the sum of individual product trends. To get the number that was rolled up as the single growth rate of IBM, the company added together the numbers generated by mainframe sales, minicomputer sales (System 7 for example), PC sales, typewriter sales (they still existed), software sales, etc. Really we are talking about a number of different products that may have their own sales trends that when taken together formed the growth rate trend of IBM.

Chapter 4
DEFINITION OF TRENDS

We started this discussion with a simple definition of a trend. For convenience we simply defined a trend as a set of measured observations that over time tend to move in a particular direction. We now want to expand that definition to include properties of a trend and what parts of those properties will allow us to further define a trend.

Consider the structure of a trend. One key component of a trend is a behavior or collection of behaviors that are novel or unique. Another aspect is how long will the trend exist? For a trend to really be considered seriously it has to encompass an increasing number of adoptions. (Adoptions: meaning more people buying—literally and figuratively—into the idea of a trend.) Lastly, the trend needs to cause a fundamental change in behavior to have an impact. These properties can be summarized as Uniqueness, Staying Power, Increasing Adoptions and Change. With these fundamental properties we can begin to think about trends in a systematic way.

Uniqueness is a key issue. If we develop the ability to spot a trend we need to think about what makes a trend attractive. We are not interested in trends that have no meaning, or trends that are dying or trends that generate nothing of interest for most people. We are looking for new trends that can be spotted and exploited. We can easily find lots of stagnant areas, but we want to find something that has the ability and capability to grow and make changes.

Staying power is another key parameter. We want to find a trend that is sustainable so that there are multiple opportunities available for products or for capabilities to be developed. We are not looking for something that is a fad or has a small duration of interest. A classic example of a fad is the "pet rock". It is clearly in the category of a novelty or a fad. In that case someone had an interesting idea and built a clever sales and marketing campaign around the idea. But, the idea did not have staying power. It was introduced, grew for awhile and then crashed in a very short period of time. In all it lasted about six months. This is the classic example of a fad. In the case of digital electronics, there is a multi-decade trend. In the case of high frequency stock trading a second could define a long term trend.

If we are to discover a trend with some staying power we also want an increasing number of adoptions. We are not looking to find a trend that is flat or that has no characteristics of growth. If we are looking at the subject of trends with the idea of making money, investments or product development we want a concept that is a growing concern. We are not interested in trends that are flat or moving downward.

Lastly, we are looking for a trend that has the ability to create change. Change is the agent that will allow for multiple competitors, as well as variations and developments so that a product category or strategy can develop. This gives us the ability to spot and exploit the general capabilities of this product or service so that we can establish and create a successful product or business.

We have discussed the idea of a trend and its potential properties in the previous paragraphs. At this point it is important to define a trend in the context of corporate or product development. So now we would define a trend as a unique and measurable set of events that exist over an extended time frame with increasing adoption by users that creates a change in the corporate or product space.

An example of a current and important trend is how fast computer technology is moving from the desktop to the mobile space with a variety of devices ranging from smart phones to tablets. This is a continuation of the trend when computers began life in closed locked air conditioned rooms and now have moved into the mainstream of personal use; almost to the comic book adaptation of the Dick Tracy watch (Samsung has recently released a "smart watch"). It's amazing to observe how rapidly consumers have embraced mobile technology to the point where most people consider it the ultimate "must have" product. When people camp out overnight to get the latest mobile device you know your trend has arrived.

Chapter 5
Visualization of Trends

There are many ways to visualize a trend. I prefer a simple graph. I prefer a graph because I can plot the values associated with how many people adopt the technology as a function of time. This gives me a visualization of the change in the number of people involved in the trend. As a simple example, I could graph the number of users of Facebook by year (or month) and I would have a representation of the total number of people who were using that form of social media. This is important for a number of reasons that we'll discuss later.

If I am an investor, I do not want to invest in a product or company that is not growing!

If I want to develop a new product or company one way to approach the problem is to create products that piggyback on top of a rapidly growing market segment.

If I am a user and want to select a social media network for talking and networking with my friends, I want to select a viable network. I do not want to waste my time setting up and building an online presence and put up content on my page if the social networking company is going to fail. If I had started a page on Myspace for example, I would have spent time and energy migrating to Facebook when Myspace morphed into a site more oriented towards business. And, as Facebook has developed, they have had to embrace new ideas and features (some of which make its users cranky). But a lot of this is Facebook reacting to the incredible fast-moving trend towards mobile platforms. They don't want to be left out or distanced from their users or their advertisers who want to reach their users on these platforms.

One of the problems with graphing information is that data often does not have consistency. So you may end up with a plot that looks like a scatter gram and you will need to figure out if there is a trend or you're just looking at random points. Generally, you can discern a general pattern or trend, but it may be difficult. Another issue about data is that it does not consist of points, but consists of ranges. So you may want to graph the data either as ranges or as an envelope around a center point. All of this is to bring visual focus to the data to determine how and if a trend is forming.

My bias is to develop plotted information on a semi-log graph. The reason for this is that a semi-log graph gives me the slope (rate of change) of a trend visually. Other people like to add features to their graphs such as moving averages. I prefer a clean semi-log graph. A simple semi-log graph is illustrated in **Figure 1**.

VISUALIZATION OF TRENDS

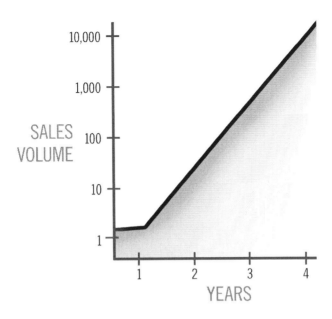

Figure 1 Illustration of a simple semi-log graph

The goal in plotting a trend is to illustrate the key features of the trend. If we are plotting a trend we want to see the aspects of increasing adoption as well as staying power. If we have picked a trend that is unique and can create change we can graphically see a trend that should interest us. One way to measure uniqueness and change is to take trends in similar areas and put them on the same graph. We can then determine if our primary trend is drawing people away (because of its uniqueness or ability to cause change) from the related trends. One way to do this is to take several products where we see trends develop and measure their market share over time.

One of the greatest trends that we can visualize is the trend in semiconductor density. **Figure 2** illustrates the growth in semiconductor density as measured by the number of transistors making up a microprocessor. This graph is a simplified version of more complex graphs that can be found illustrating Moore's Law which states that transistor counts on a chip essentially double every 18 to 24 months.

Different people have slight variations on this effect but the graph shows the basic trend. Whether this trend can continue and for how long is a constant subject of debate. But with quantum computing and nanotechnology computers no one can tell where we are going over the long term. However, it has been quite counterproductive for people to bet against the trend expounded by Moore's Law.

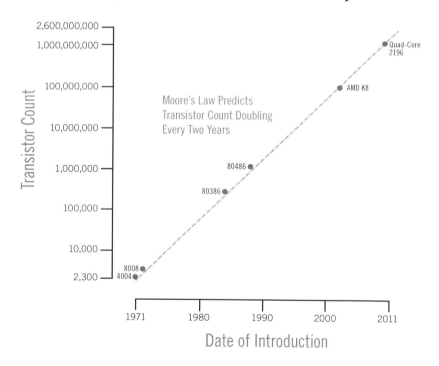

Figure 2 Illustration of the growth of Transistor Density on Single Chips over Time—Moore's Law (Source: Intel)

CHAPTER 6
PROBLEMS WITH TRENDS

Duration is the key issue with a trend. How do we determine if a trend has the ability to continue? And, can the trend not only continue, but continue at its current rate of increase or even accelerate its rate of increase? These are important questions and the answers will determine if we can exploit a trend to gain a product or corporate foothold.

In this case, we are not necessarily the generator of the technology that we are trying to exploit. We are just trying to determine when a trend has legs and can actually cause a change in consumer behavior.

A big problem is the fad. The difference between a fad and a trend is usually duration. If something is only going to last for a year or so you will not have time to recognize and exploit the result. Thus, I try to look for trends that are I think have staying power. I look for trends that look like they continue for multiple years. If a trend seems to have a large component of novelty associated with it, I would be careful and try to decide if it was a fad. Sometimes, something will start out growing rapidly and then it will collapse. Such fads can be deceiving and they can look like a trend. Take the "Pet Rock" for example. The idea was that if you owned a pet rock then you essentially had the equivalent of a pet. However, unlike a real dog you would not have to deal with things like walks and poop patrol. The idea caught on quickly and bunches were sold, but just as quickly the market collapsed. There was just no real market or value compared to owning a real pet. They were a nice conversation piece but at the end of the day they were essentially dust collectors. Other examples of fads are easy to spot and understand. One fad that seems to come around every summer is that catchy summer song that gets played over and over again on the radio creating the summers one hit wonder.

Sometimes trends start, fail, re-ignite, fail and then finally get on track. If you look back at the phone industry, you see the early ideas of PDAs and phones leading (after many false starts) to the development and evolution of the smart phone. But, there were many attempts and failures along the way. If you were an early pioneer developer and manufacturer you could have lost a lot of money before the right technologies finally came along and generated the consumer market that made the smart phone a real consumer product category. It took a confluence of several trends to actually make this happen as the size of cell phones, batteries and a host of other technologies had to all come together to put us where we are today. Just think back on the early Motorola cell phones that looked like a brick and had about an hour's battery capability. Then we also had PDAs that had very specialized and highly integrated (read closed) operating systems and applications.

There are many problems with trends. Not the least is that it is very simple for someone to take two points, draw a straight line between them and claim that a trend has formed. Then they compound it by extrapolating the trend line into the future. The sad thing about this type of simple analysis is that it is all too common and it is misleading. And, worst of all it is hazardous to your financial health! It does not represent reality. The most classic case of this that I encountered was with a group of people who were nearing retirement. At that point they had a few years left before retirement and were planning their retirement finances. They had the good fortune to have seen a set of consecutive years of approximately 15%

compounded annual growth in the S&P 500 index. They were fully invested in an S&P Index Fund. They were all using charts generated by a financial advisor that showed quarterly results of the S&P 500. This chart showed a nicely increasing line and the future retirees had extension lines on a chart showing how much money they would have in future years using the same assumption of 15% compound growth. In talking to them it was clear that they did not understand that the reality was that the trend had been smoothed by only showing quarterly data and it was long in the tooth. It had been going on a long time. The length of time was a serious problem. An index of that nature with that type of "long" term trend was way out of the norm. Needless to say many of the people retired on the basis of this trend, right before it hit the wall and collapsed. Unfortunately for them, they were really surprised at what happened! Hopefully they did not spend all of their money in one place or at the start of retirement. But this type of incident is all too common. Particularly in the financial markets, such examples are common. One only has to look at recent examples like the Internet bubble, solar energy, ethanol production and at times various commodities such as gold.

Some other problems with trends are just basic human nature. People have a tendency to believe that "this time it is different", or not understand that once a trend has formed lots of other people will also find the trend. People will always take the easy way out and once a trend has been spotted it will eventually be overwhelmed by hangers-on. Modifications are likely to occur due to an emerging or substitute technology taking the trend in a radically different direction. The latter can cause multiple spin-out trends with different characteristics to form.

In my previous books I was clear to point out that there are a lot of smart people. No matter how smart you think you are there are other people who are as smart or smarter. If you have spotted a trend, you can expect that someone else has also spotted the same trend. In fact, they may have spotted it earlier. No matter, we can still do well, but every minute we spend thinking that we have something that no one else has, the greater the danger. Eventually others will figure out the same trend. If we are using the trend to develop a new product, the second we introduce the product everyone has the trend figured out as they will analyze the product and figure out its basic origins.

It is never as different as you think and eventually (as you are successful) you have to show people what you have done and then success breeds copy cats.

If nothing else, you stand a chance of your own people trying to break off and copy your results. If you are very successful you will find an incredible number of variations around your trend develop. That is why companies like Microsoft and Google end up being the hotbed of spin-outs. Microsoft rides/creates a trend and as people see the trend they find smaller trends they can exploit and break-outs occur of new concepts that may be viable enough to start a new trend.

Chapter 7
Decay of Trends

If a trend is strong and runs for a long time, it will eventually start to run into problems created by its success!

Variations will develop, the trend will have to reinvent itself, competition will develop and the trend will run its course. That does not mean that it will be a disaster, but at a minimum it will mature.

Consider the computer industry. At the start, people (the famous quote that the world or the U.S. would need just six electronic computers has variously been ascribed in Wikipedia to Howard Aiken, American computer pioneer, Tom Watson Sr. who brought IBM into the modern computer age or Charles Darwin in 1946, grandson of the famous naturalist) thought there would be a very few computers in the world. Some people actually thought that the total numbers of computers needed by the world could be as few as 10 or 20 units. Regardless of who said it, they were wrong. But, look at the trend. When it became clear that the number of computers needed would/could be a large number we get the mainframe business.

TRENDS

Figure 3 Mainframe Centric Data Processing Center circa 1965

Once we had major corporations building mainframes, the technology and numbers of mainframes sprouted like wildfire. One company dominated, IBM. But there were lots of others that were in the business: Burroughs, Univac, NCR, Control Data and Honeywell (collectively known as the BUNCH). But they were not the only builders of mainframes. We also had specialty manufacturers: GE (time sharing systems—later bought by Honeywell), SDS (scientific computers) and Amdahl (plug compatible IBM computers). The mainframe business took off and sub-trends sprouted.

Software became a trend spawning companies like CSC, SDC and Analysts International. We could give a litany of software companies and what they did, but we need to understand that like mainframe companies they had their day and there were bunches of them in business at various times.

And, because of expense the minicomputer market developed. Then the minicomputer market started a new trend and it went through a great growth phase before it ran out of steam and its trend sputtered.

But, do not fear just as the minicomputer faded the personal computer (PC) revolution began and the trend and cycle starts again. Products and companies come and go. New markets are developed and die off.

But, do not despair because now we have the smart phone business and we get tablets. See how this works.

We not only have tablets but we have mini-tablets and a new software industry known as apps.

So by now you know where this is all headed.

Trends will form and they will grow. If successful they will spin out more trends and eventually some part of them will collapse and some will morph into other trends.

The key is that such change will come sooner than you think and will probably take forms that you have not thought about. People playing the trend game will try to position themselves to maximize their results from having spotted the trend in the first place.

Chapter 8
CPT—Death of a Trend

Trends can die. One of the best examples is the stand-alone word processor. At one time most documents were created using a typewriter or some form of a typesetter. A real advancement was the stand-alone word processor. One of the best examples of this technology was from CPT Corporation founded in 1971 in Minneapolis.

CPT stood for Cassette Powered Typing. It was the nomenclature of both the company's name and its products. Essentially the product was an attachment that could hold some audio cassettes and connect to a Selectric typewriter. The user would type and the cassettes would record and allow the user to modify text. When the user was done they could print the document and save the information on a cassette. This was really a great idea as it made documents easy to correct and simple things like a customized form letter could be created and stored for future use.

CPT was headquartered in the same town where I lived and friends of mine were early adopters of the technology. So I got to see it in use and understand its advantages up close.

As technology changed so did CPT. They moved up to microprocessor-based solutions that contained graphic display technology and storage on floppy disks.

TRENDS

CPT was a great company with a great product and its sales showed. It had a well-established sales trend that was growing fast and products that were evolving nicely. The model that I remember best (the 8000 series) was near the end of their product cycle. CPT was selling a full featured word processor that cost about $16,000 dollars. It had a great display and used the early Intel 8080 processor with floppy disk technology. This machine could literally stand up and produce your typical office documents at an incredible rate for the time. CPT was rewarded for this with a very rapid growth rate. It had competitors but it was really the only pure play word processor company.

When I cofounded the first business I started, I really wanted one of those CPT machines, but could not afford it. Instead for document processing we used Correcting Selectric typewriters from IBM.

Figure 4 CPT 8000 Series

These worked just fine but we would have to cut and paste together reports and complex documents. Eventually we developed some simple techniques that would allow us to use the early Apple IIe computer to automate our work-flow. Then, the IBM PC came out with a very nice word processing package.

At this point you can stick a fork in CPT and declare it road kill. We can use CPT as an example of the death of a trend, so let's examine the details of how they died.

CPT had a single function machine. It was great at what it did. It booted fast, had a screen that let you see the document you were working on in page format with black text on a white background. It allowed you to store your document and drafts as you went along. All of this for just about $16,000. Even better it was a high-flying growth stock. And, it had survived the first real technology challenge, moving from cassettes and Selectric technology to microprocessor technology.

Compare that to the IBM PC. This system cost about $5,000 and came with some simple applications. It had a horizontal screen with green letters on a "black" background. One of its simple applications was word processing. So conceptually you could do the same thing on a PC that you could do on the CPT machine. But, you had to work harder. If you took on this challenge you got some extra capabilities for your trouble and you saved $11,000. But was this enough to cause you to go the PC route over the CPT machine? In retrospect, we know the answer.

We will examine the death of the trend.

CPT provided a clear tradeoff. You get a sophisticated and elegant single purpose machine instead of a less elegant but highly functional general purpose machine, but you pay an extra $11,000.

CPT tried to counter this (threat to its trend) by introducing a lower priced version of its machine. This machine had more power and was priced around $10,000. But this only slowed the death of the trend.

The numbers tell the story. If you have a growth rate of 60% per year and you have to cut your prices by about 40% to keep your revenue base you must sell 60% more the next year. Let's look at the math. If you sell one machine you get $16,000. To grow 60% the next year you have to reach sales of $25,600. If you cut your price to $10,000, then you need an additional $6,000 to breakeven. And then you need an additional $9,600 to accomplish your projected growth. At the price of $10,000 you need to sell 2.5 times as many machines as you did before to get your expected growth.

You would be lucky if the change of technology to PC-based systems would only hurt your pricing to $10,000. Like I said, customers faced with this tradeoff put the fork in CPT and the trend broke. It broke big and quick. CPT became a footnote in history even though at one point it was considered a magnificent growth play.

You need to worry about the duration of a trend. Once a trend becomes well defined, and if it's financially successful, it will quickly draw lots of competitors many of whom may be better managed or financed than you. In that case your involvement with the trend is over and you need to move on. Or, you can adapt by raising your own massive funds or management team.

Many people argue that money is difficult to get for new start up companies. Yet it has been my experience that the world is flooded with money for startups at the right stage. But as a successful founder your problem will be to decide the terms under which you accept people who want to get aboard your trend. If someone wants to get aboard you need to understand that they are not doing it for you they are doing it for themselves and their investors. You just happen to be a convenient vehicle. Further, their institutional processes may not mesh with your desires, but like everything in life it is a negotiation. You have to decide what you will put up with to carry your successful trend forward.

The computer industry is a great example of what can happen even in a simple trend. First there were mainframes. Then, along came minicomputers. At some point, supercomputers jump into the fray. Then we add personal computers, tablets and mini-tablets.

Each of these developments creates a trend or set of trends. A new set of industries come and go. The creation of a new trend, the spin out of a trend or the decay of an existing trend all offer chances for you to participate in the ramifications of the identified trend.

CHAPTER 9

VARIATIONS ON A TREND

The quickest way to survive and prosper if you create or spot a trend to exploit is to study the trend and then build a variation of the trend. You do not have to re-invent the wheel; you just need to re-purpose the wheel. You can use the basic wheel and just use it in a different and unique way. You can also reinvent the wheel if you want. In this case you might want to put in different spokes or use a different material. A clear example of re-purposing of a wheel was creating automobile wheels that were not just made with a chrome coating but contained a shell that would pick up energy from the motion of the wheel and spin even when the car was stopped. This created the illusion of motion of the wheels at all times even when the vehicle was stopped.

If you have a successful product that looks like it is setting up a new market and there will be a sustainable financial flow of significant size you will end up with a lot of competitors. The venture capital industry is not really set up to innovate. They are however good at flooding the market with copycat or variations of successful products and letting their financial muscle sort out the winners and losers.

The basic problem with a trend is that you generally spot it after it is well formed. But, it is possible to take a trend and enhance it. If you are not on the creative side of an industry you can still profit as you spot a trend. The solution to understanding and profiting from a trend is to realize that once a trend forms you still have available to you variations of the trend as well as short term trends that may form inside a basic long term trend.

Particularly in the technology industry, you can easily see that trends form and variations become common place. As a simple example consider the smart phone or the tablet industry. For years designers thought that the basis of mobility in computing would be formed around a PDA (Programmable Digital Assistant) and people tried to graft some level of phone functionality onto the PDA. Although these efforts were marginally successful, Apple's introduction of the iPhone left companies caught flat footed. A whole new industry was formed in hardware and software, smart phones and apps respectively. The introduction of the iPad furthered the push towards mobility. At about the same time E-Readers, which are really just special purpose mobile devices, took off.

With companies approaching these products from different angles, a variety of devices established a new trend for mobile computing that is just really starting to be felt. Some applications are pretty simple like letting a merchant use a cell phone as a credit card point of sale device.

Once you think about how mobile devices can be used you understand that mobility is a new and rapidly evolving area where a number of trends and spin outs from the original trends will form.

You can have many variations of the trend. Companies like Facebook try to monetize their information about their participants onto mobile platforms. Where this will go is difficult to predict and understand. However, because of the stock markets, even if you do not create trends you have the ability to make money by identifying trends.

This is why companies, investors and individuals work hard at trying to spot trends. If you spot a trend early enough you have an opportunity for making money or even creating a company in a similar business with a product that is a variant of an existing product.

One caveat, however. Keep in mind if the product you are targeting is highly successful you will find a very crowded space.

As it becomes simpler and simpler to build a new product it becomes more difficult to dominate a market. Consider the realities of the smart phone or tablet markets. Today it would be very difficult to enter such a market unless you have big connections and capabilities to manufacture the product. Yet it would be quite simple to develop an app and sell it through the app market places. And, in fact, you could develop apps that would replace many standalone pieces of equipment and this is occurring today.

Even the functions of the Kindle book reader are available for your iPad. Some functions just appeared on smart phones as the phones evolved. This leads to questions about products. For example, is the Maps function on your smart phone good enough that you do not need a GPS system? The technology of apps is making a major contribution to people's ability to be mobile and still connected.

CHAPTER 10

REINVENTION OF TRENDS

I am biased toward reinvention. I believe that progress is made incrementally. Chairman Mao had a different approach. He believed in the "Great Leap Forward" or GLF. I am not opposed to people shooting for the moon and trying to develop a GLF. However, personally I think that idea has a low probability of success.

I favor the idea of looking for trends. And when you can identify them try to figure out how to reinvent (or how others would reinvent) or cause the trend to be modified.

In a sense this is a simple process. Again I look to the computer industry as an example. Regardless of where you enter the computer market you need hardware. And, you need to figure out how to defeat the competition. You need to build a better product. How you do that is pretty straight forward. At one level you can change aspects of the hardware. Make it cheaper. Make it faster. Make it smaller (lighter weight). Make it easier to use, etc.

TRENDS

Figure 5 Chairman Mao

There was no GLF strategy in the computer market. You can trace the development of the smart phone and tablet back to the mainframe in a series of straight forward functional changes in the computer industry. You can also see how trends get reinvented by simply changing a few basic capabilities of a product.

But, compare the economic results of China to the computer industry. How do they compare on something like total products shipped? What about growth rate? You get the picture. Even though China is now an economic powerhouse, how much progress have they made compared to the computer industry? Maybe they are becoming an economic powerhouse because they have conveniently set aside any more Great Leap Forward plans.

Even more telling, you should compare the progress of the computer industry to the progress of China during Mao's five year GLF. Like my teachers used to say in math class at the end of a proof, QED. (QED, Quod Erat Demonstrandum, basically translates from Latin as hence proved or demonstrated.) Since I have not provided a rigorous proof, I just ask you to think about the information and look at it in the sense of the trends generated by these strategies. Also, I handicapped the reinvention idea by illustrating it with the computer industry, the most cost-competitive and cutthroat industry that humanity has ever known.

Progress is incremental. Reinvention pays off. Spot a trend and figure out how it can change or evolve.

Concluding Remarks on Trends

Simply said, if you can spot a trend, the trend will be your friend. It is not always true that you can spot a trend but if you do, you have the ability to adjust your strategy and create ways to position a business or product. If you are lucky enough to spot a trend associated with product sales, you can position a new or modified product that will put you in position to develop a new company or create a competing product.

Generally, trends will tend to continue until something disruptive creates decay or changes the focus of the trend. This can be easily understood by thinking about how cars drive. A car will tend to go where you are looking. This is why texting while driving is so dangerous because you have no idea where the car will go during the time you take your eyes off the road. (Someone once said it's the equivalent of driving with your eyes closed.)

If you can spot a trend, you should assume that the most logical path of a trend is in the direction it is going. Fighting a trend is generally counterproductive. Yet a number of people have been quite successful in fighting trends particularly so in the field of investing where hedge funds try to figure out when a trend is about to break and reverse course. But, spotting these inflection points is very difficult and risky. Being risk adverse I try to avoid fighting trends.

PART 2
WAVES

Introduction to Waves

Previously we examined the idea of a trend. In the following chapters we'll look at waves and how they derive or spin out of a trend.

Waves are important because they are the result of a trend. Once people can identify a trend they try to capitalize on the trend. One good way to capitalize on a trend is to develop a product or company that exploits the trend. Although I make that statement glibly and easily, the actual practice of building a company or product is very difficult. Building something is fraught with risks; but, for people who can spot a trend in its early stages and ride the actual wave the results can be spectacular. Unfortunately, your ability to spot a trend before anyone else is probably limited because there are lots of other smart people looking to spot a trend also.

By itself a trend is not a viable way to develop a product. It is the waves that a trend can spawn that are the basis of product development. Consider the simple development of the transistor. The development of the transistor created a trend of developing and implementing smaller, denser, more powerful and more reliable electronic devices. The building blocks (transistors, integrated circuits, microprocessors and programmable logic arrays—to name a few developments of note) that form these devices came in waves of technology innovation. And, you did not (and even today do not) have to be a pioneer in the electronics business to "profit" from the waves of technology innovation spun out of the transistor revolution.

The question is whether you can spot when a trend causes a wave of technology innovation to form (and possibly destroy a previous wave) and begin to take off in a product sense. A trend can cause many waves. The trend to high density digital circuits has spawned many waves of technology products including mainframe and tablet computers. A trend is a long term set of events that develop fundamental capabilities. A wave is a set of products that spin out of a trend. Unfortunately, it may take a long time for a wave to gestate. But a strong technology trend will probably set up a series of waves. And, you probably cannot foresee all of the variations that will derive from the basic technology. Again, look at the development and manufacture of the first transistor and one of its resultant products— the transistor radio.

In the mid-fifties I bought my first transistor-based product, a radio, which had six or seven transistors. It was a great product and I loved it. Because of its size and the fact that it was battery powered I could carry it around and have access to the media of AM radio; but, let's be real. Did anyone envision that we would walking around (50 years later) with the type of portable sound devices and computer power that we have today in our smart phones and tablets? Or, did anyone envision the path of change and the waves of innovation that would drive the semiconductor industry from the single transistor to the current chip technology that is generating the end user devices we see today.

You might be able to see the next wave forming, but seeing what works as a product strategy two or three successive waves from now is very difficult. The problem is that once a wave forms and takes off people see how the wave is forming and they produce a set of products off that wave. New groups trying to enter the market must naturally try to build products with features that will get them a unique competitive position. Product waves tend to build off of successful product strategies and, even if we have a clear insight into a current product wave, we do not really know how it will evolve into the next wave that we could eventually identify.

If spotting waves as they form were easy, everyone would do it successfully. It is not easy, but the discussions in this book will give insight into the process of spotting and exploiting opportunities formed by trends, waves and windows. A further complication is that a really successful product, a big wave product, may need multiple technology trends/waves to converge at the same time. The semiconductor revolution begets the computer revolution. Natural progression of the semiconductor technology combined with operating system and disk drive technology begets the personal computer. However, you need more than a simple evolution to get tablets. You have to have the portable personal computer/laptop revolution combined with extremely powerful chips, the internet, telecommunications in the form of Wi-Fi and 3G/4G technology, a new concept of application software (apps and the app stores) as well as lightweight and powerful battery technology. Then, you must get consumer acceptance in such volume you can drive the price down to the levels that the product is considered a simple consumer device. If you look at the mobile devices of today it is difficult to think that they started back with the Osborne/1, the first portable (luggable) personal computer sold in the early 1980s. Consider what happened to the Osborne Computer Company—it went exponential in growth and then dove directly into bankruptcy!

You can start to see the evolutionary path. Advances in semiconductor technology push the creation of new groups and new products. Waves and trends begin to form. You know that major developments are just waiting in the wings.

We will discuss these issues in depth in the following chapters 11–20.

Chapter 11
Wave Concepts

The concept of a wave is pretty simple. It consists of a set of developments that are derived from a trend and from a product perspective the trend can be exploited. In the ocean we have waves forming as swells move onto the beach. The storms that push these swells towards the beach are our trends. You do not have to generate the big wave or product cycle to profit. You just have to recognize that the trend is forming a big wave and you can act on the trend (or some part of it) from an investment, product or company formation perspective.

The fundamental building block of new technology is the work coming out of research and development labs. It is within the lab that ideas are generated and brought to different levels of maturity. If enough technology comes together, new products can be formed and if developers can get early adopters hooked a product trend can evolve out of the basic technology. But a new product concept may need to take advantage of several trends. Even if you have a trend of semiconductor chip density you will not get portable computers unless you also get new battery technologies that decrease the weight to where you could carry the computer.

Think of these labs as a proving ground for concepts and potential trends that get launched to the shore and try to form a wave. The further away the trend is from forming a wave, the riskier your bet on that trend.

This is where it is difficult to figure out whether we have a trend—does the trend's direction toward the shore make sense and will it form a wave. These swells are, in a research and development sense, ideas that can become the basis of a new product. The further out the idea, the more work that must be done to create a trend that generates a wave. Consumer demand creates the wave. The priority is to establish buzz among early adopters so that the waves begin to form and push onto the shore. As this happens, competition will develop and, if it's meant to, a big wave will form. If you have actually spotted a big wave, its rapid growth probably will overwhelm your capacity to ride the wave into a successful product and company. The reality is that there is so much money in the world chasing high returns, that a trend that converts to a product wave will attract tons of money. I know there is high unemployment, the middle class has lots of pressure on it, etc. but the reality is that for people with good ideas there is always a way to profit and ride new product concepts. Look at the growth of the economy and unemployment rate during the rapid growth of Apple due to the iPod, iPad and iPhone. Was there a recession or depression at Apple or its suppliers?

Fundamentally there are waves forming in the ocean at all times. There are always new trends. If a research and development lab can create a disruptive trend and the lab can cause early adopters to use a product and create a buzz, then the new technology gets launched to the shore as a wave.

In technology, the disruptive trend creates the buzz with early adopters, which creates the pressure for a technology product big wave. Big waves present opportunity! But, you can profit even from little waves.

The technology trend we want is rapid user adoption, which causes rapid price drops. The faster the price drops, the more users will adopt the product. When you get a product that quickly moves from early adopters to mainstream consumers you will see a rapid and sustained investment to eliminate the competition before it can gain a serious market foothold. For example, the pocket calculator, cell phone, personal computer, and modern television markets all developed very quickly, and the resultant waves took on steep characteristics as things moved rapidly to a price shootout and rapid product churn.

    Investors recognize a trend is forming and jump on it when rising user demand creates a wave. If a trend looks like it's forming, the investors will tend to jump into the fray. This is where the correct identification and exploitation of a trend can allow you to profit even if you are not an inventor or product developer.

Chapter 12
IMPORTANCE OF WAVES

Waves are important. They evolved as the result of an evolutionary technology trend that has created a product opportunity. Product opportunities do not just drop from the sky. They are the result of years of effort and positioning that allows the end user to be able to buy a new (and hopefully useful) product. When a product becomes recognized as useful the product wave has formed and you have a chance to build a product, company or wealth within a short period of time.

Trends can create multiple waves of opportunity that will span a number of waves. This allows an entrepreneur multiple entry points to create their great invention.

Waves of technology moving into the product space create havoc for existing companies. A company that has an established market position can be eliminated if they get stuck on a particular wave and do not make the jump to the next wave.

If you are able to spot a trend and then identify a wave you have a chance at putting your product in play and catching a portion of the market. But the reality (and importance of waves) is that once a trend has formed the trend can cause many waves over an extended period of time and it is very difficult for product designers to change with each new wave of technology. Like the general population, designers become creatures of habit and over time they tend to grow more and more conservative in their designs. It is a very rare designer that can change and succeed for more than a couple of design cycles and thus be a significant factor in multiple technology waves.

If you recognize that trends evolve, then you should also realize that different waves can spring from the trend. You might be able to identify different wave structures so that you can evolve your product and its position over time.

You will also face several problems associated with waves. When you are on a beach and look at waves coming onto shore you will notice that they are all a little different. Waves differ in many ways. But one of the key issues is the duration of a wave. Waves seem to come in groups and may have slightly different properties. Waves of product cycles share similar properties. So it is critically important to spot when a wave is forming and when it is starting to decay.

If a new set of products has just been introduced you do not want to introduce a "me too" product.

You need to figure out the wave structure. See if you can position your product with features that make the other products look dated while yours is perceived as the new "state of the art" product. A perfect example was detailed in the December 2 2013 issue of Time magazine. The popular mobile game called *Candy Crush* really took off after they made some design changes like making it playable in both portrait and landscape orientation. According to the article, this allowed users to play it with one hand favored by subway commuters and "office workers holding their phone beneath a conference table where they could continue to play the game." They also synched it across multiple platforms. So users can play it offline or online and their progress is saved once they have service again. That means you can play the game anywhere, anytime. Minor changes perhaps but to date the game has been downloaded over 500 million times and played more than 150 Billion times! The article concludes by saying that the developers know they have a hit but they also realize that it's a constant battle to stay ahead and they must port the *Candy* magic to other titles.[1]

So what waves and the Candy story illustrate, is the constant change and nuance of continual product evolution. Our challenge is to take that very important observation and figure out how to position a solution to a problem or to create a new product that becomes the next big thing and creates a new product category.

---

1. Eliana Dockterman, *Architects of Addiction*. *Time* Magazine. 2 December 2013

Chapter 13
Wave Types

In a simple world, waves can be characterized by two parameters: duration and size. However, we live in a complex business world and waves can have other attributes particularly in the business world. For example, in the ocean a big wave will tend to have not only a large size (height) but it will have a fairly long duration. In the business world waves can be more complex. They can have wildly varying properties.

 As an example of this phenomenon, the pocket calculator created a big wave at its introduction but the wave was more like a spike than a slow moving big wave as competitors formed at an amazing pace. The price drop was so rapid that the market flattened out. The future changes to the technology made it seem like a spike followed by a set of very small waves moving along at a slow pace. The PC industry and the office automation software industry had a similar profile. In fact, the office automation software for PCs quickly moved from a business of selling software to a business whose main strategy was to sell software upgrades as the number of new products sold was quickly dwarfed by replacement software.

 In business you can have more complex wave structures than in nature. Basic technologies along with money and competition combine to form a start point for products and services that create and form waves. Further, you can have spin off waves that are generated by unintended uses of the basic technology. So, technology wave structures can be very unique and unusual. This unique wave forms when its basic structure gets modified by the forces of competition and money.

In the business world an example of a large wave is the rapid growth of the computer industry. It has had several sizable sub-waves within its structure notably the mainframe market which has morphed into the server market and then into the cloud market; while the minicomputer market morphed into the PC/microcomputer market. There is also a completely separate set of waves associated with the industrial and process control market as well as the market defined by high reliability computers for military and space uses.

In the business environment shrewd observers know that money follows technology because they know that it's difficult to conceive, create, prototype and produce a new product. However, it is relatively easy to copy a product or prototype once you have seen the basic concept. If you have seen a pioneering product and it seems to be getting a lot of traction and the market has seriously large potential, it is common for large amounts of money to flood the market. And, the money and competition will come quickly. Very often the first mover (developer of the initial technology concept) will probably lose out. The first mover may be stuck with an obsolete version of a potential rapid growth technology because while they were spending time on the actual conceptualization, development and implementation of the technology, another competitor seized and improved the basic technology and raced past them.

In the business world a wave can abruptly end as another competing technology can overrun and make an existing technology obsolete. Yet some technologies will go on for a long time. The problem with technology is easily illustrated by the problems being faced by archivists. They try to make sure that they have technologies that allow for long-term preservation of film, audio files and pictures that have been taken and exist in a wide variety of formats, some of which are technologically obsolete. This is but one example of how a short-term wave creates long-term disruptive problems for a small group of people interested in capturing and preserving historical materials.

Chapter 14
WAVE DEFINITIONS

The concept of a technology wave is not very difficult to explain conceptually but it is very difficult to recognize, understand and exploit.

A technology wave is simply the process (and observation of the effects) of a widespread and rapid dispersion of a technology. In general, people and institutions are resistant to change, but sometimes a technology emerges that forces massive change in the way people perceive and operate in their day to day lives. If the technology change has a dramatic impact on productivity it will be adopted rapidly and big changes will occur in a very short time span. This is what is called a big technology wave.

Usually such waves are due to several factors. There are many factors that could cause the wave to form, but some of the common factors are: productivity, cost or process improvement.

Productivity is one of the easiest ways to cause a wave to form. If you can find a technology that changes the process of a factory and thus increases the productivity of the factory you can create a wave. In terms of changes in the ways factories are run, any improvement in the productivity of the factory is always on the mind of the factory managers because productivity increases translate directly to lower unit costs. This constant need to drive towards lower unit costs make factories a prime target for new technology innovation and thus the formation of waves. Some waves, like the advent of low cost and highly precise robot controls, can cause a factory to have major productivity changes and thus provide the potential for dramatically lower unit costs. Take, for example, the 2012 purchase of Kiva Systems Inc., a robotics company, by Amazon.com Inc. By some accounts it's estimated that Kiva's automated systems could shave 20-40% off the $3.50-$3.75 cost of a typical Amazon order. It's projected that once all savings are fully implemented across Amazon's vast warehouse system the cost savings could reach $900 million a year.[2]

It is no surprise then that as factories automate they will need less labor and manufacturing jobs will disappear. The drive for lower costs does have an interesting side effect. It may require less labor, but the type of labor also changes and the need for highly skilled technicians becomes the dominant required labor. It's pretty easy to understand this phenomenon. Consider the employment in the auto industry in Detroit. Not only is less labor required, but the quality of the produced vehicle has increased. And, the type of labor force has changed dramatically.

Another factor that can cause a wave to form is cost. If you can deploy a technology that will dramatically change the cost equation, you can form a wave and depending on the perception of need you can create or replace an existing technology.

Take the computer industry. Today computers are everywhere in our daily life; this was not always so. In the beginning they were in closed rooms available to only a few people. However, as they became cheaper to make, new variations began to change the cost equation and they begin to proliferate in waves of technology. Mainframes spawned minis. Minis spawned micros and personal computers. Each iteration was defined by lower costs. New price points became prominent and the devices became easier to use. As costs dropped (and functions increased) the computer moved from being a luxury item to a daily must have and the uses of the device

---

2. Greg Bensinger, *Before Amazon's Drones Come the Robots*. WSJ.com. 8 December 2013

became prominent in most people's lives. If you can make a technology easy to use and get its cost to a consumer level, large markets will become viable product opportunities.

Process improvement is another common wave form. Look at the evolution of the personal computer to a tablet. Some may argue that a tablet is actually a new category, but functionally it is a very close relative of a netbook which is related to a laptop which is really a portable PC. In this case, better battery technology and less weight combined with different capabilities and functions to effectively create a wide variety of product categories. Effectively the combination of a smart phone and portable PC is now the market category of tablets.

The effect of these wide varieties of changes also changes user's perception of need. Advances in both hardware (integrated circuits, batteries, etc) and software (apps, office productivity tools, etc) coupled with new communications technologies has resulted in waves of technology that have fundamentally changed the way people work, play and relate to each other.

When we look for a wave we are just looking to find a technology driver that will cause the adoption of new technologies. If the technology disperses rapidly into consumer markets then we have the chance that the wave will be a big wave and real money can be made and/or lost by its adoption.

Chapter 15
Visualization of Waves

After a wave has formed it is easy to see its effects.

Spotting a wave as it starts is not easy!

Visualizing that a wave is forming is very difficult.

In the technology field, you either catch a product cycle or you wait until the next cycle starts. In some cases, the waves and technology product cycles may be pretty small. Far out in the ocean there are swells. These swells are similar to concepts that begin to form in a research and product development lab. As the swells reach the shore and the products begin to develop, demand builds and the swells form waves of different sizes. The weather that pushes the swells into shore as well as the beach geography determines the resultant wave size. Weather corresponds to demand creation.

The wave moves onto shore as the pressure of consumer demand moves the wave to full height. The wave breaks when other manufacturers enter the market, and the prices disintegrate as the wave crashes onto the shore. Lots of waves crash into the shore with no riders. Similarly lots of technologies that might ignite massive change never reach fruition. Reasons for failure to launch vary. The reason could simply be that no one caught that particular wave and the wave became obsolete just as another incoming technology took off. Yet, even as we start down the face of a wave, and catch that new technology, it doesn't mean that it will become a solid product technology. The technology may crash and burn or it may be lost to another new and innovative technology before it becomes a serious market force.

Technology follows cycles like waves and swells. Different technologies mature at different rates. Like a big wave surfer, you must learn how to surf and develop technologies. You do not start out conceiving new products, just like you do not start your surfing career on 60-foot waves. You are committed once you launch onto the big wave product cycle and try to turn that technology into a rapidly growing product and company. You cannot turn back; you must ride the wave out, or you will wipe out. With practice, you will learn how to surf big waves.

There is always an opportunity, and you can surf on a lot of beaches. However, there are few legendary surfers in technology and ocean big wave surfing. Two of the best known are Laird Hamilton (big wave surfer) and Bill Gates (technology big wave surfer). Both are legendary figures in their respective disciplines. Yet, there are hundreds of successes in both fields—people who are successful but not superstars. You may have never heard of these other successful people, but there are a lot of them. Today if you stopped and asked most people on the street who is Arthur Fry you would get a blank stare. Yet Arthur Fry's invention touches thousands if not millions of people every day. Arthur Fry was the co-creator of the yellow Post-it or sticky notes which he created at 3M. You do not have to be the superstar to be a great big wave surfer!

A wave consists of a large wall of water that is headed to the shore. A number of factors determine the size of the wave, including the underlying beach structure, the prevailing winds, and the size of any storms out in the ocean. However, size is not the only issue associated with a wave structure. A further issue is the size and duration of the churn that forms as the wave breaks into white water and washes onto the shore. Undertow that pulls swimmers back into the ocean is another factor. The structure of the shore or ground underlying the structure of the water is also important. Whether there is sand or coral beneath the wave can have a large effect on the type and duration of the surf as well as the surfer's ability to survive.

The further out in the ocean you are, the more you are simply riding up and down on swells with little direction toward the shore. These swells are, in a research and development sense, ideas that can become the basis of a new product. The further out the idea, the more work that must be done to launch the swell towards the shore. However, a stiff wind from a storm can get a swell moving toward the shore and forming a wave. Consumer demand can have the same effect on a research project as the wind can have on swells. The trick is to establish buzz among early adopters so that the wave begins to push into the shore. As this happens, competition will develop and, if it's meant to, the wave will form. With enough pressure, we form a big wave.

If you are a big wave surfer, you have a serious identity issue to solve. At the very start of the product concept, you may be all by yourself or part of a small team. This is when you are trying to figure out whether a specific swell will turn into a big wave and whether you can generate a big wave, high-growth company from your product concept. At this point, you must have a wide perspective: developer, manager, marketer, and investor. Once you have correctly identified the big wave product and have received some traction on the face of the wave, you can then start to specialize the talents of the team into separate disciplines. When you are trying to spot a big wave, you must do everything. When you ride a big wave, you must get help from a variety of sources. If you have actually spotted a big wave, its rapid growth would overwhelm your ability to do everything and ride it into a successful product and company.

The growth of internet hosts during the 80s and 90s illustrates the reason for the internet bubble. If you have a rapid growth of computing hardware you can imagine how people were trying to figure out ways to exploit the technology. The trend is clear. Waves of technology solutions came onto the scene to exploit these new interconnected hosts and add value by allowing people to do all kinds of new functions and achieve productivity gains. At the same time you also end up with waves of silliness. Just think about delivering pet food over internet channels while trying to beat out your local grocery/big box store on price. It was some form of insanity. This happened because you had excess capacity and more connections that allowed new applications (of all kinds) to evolve and waves of technology opportunities appeared in search of a problem.

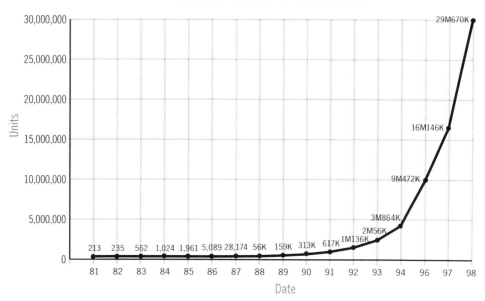

Figure 6 Illustration of the growth of internet hosts from 1981 until 1998. (Source: Coffman and Odlyzko, 'The size and growth rate of the Internet," October 2, 1998)

Chapter 16
PROBLEMS WITH WAVES

When you spot a wave you have no idea of its size or permanence. In some cases waves have very short life spans. In other cases waves appear to last for long periods of time bordering on permanence.

Figuring out if a wave is real enough to exploit is a big issue. Waves can have a short life span for a number of reasons including:

- they serve a temporary need and get shoved aside by a better technology,

- they fail to bring about an economically sustainable change to user needs or expectations or

- they are simply too expensive to make the leap to a large consumer market.

In the case of the wave being shoved aside by a better technology or serving a short term temporary need, the wave has a fundamental problem with its economic positioning. A wave can exploit a small but fleeting niche during which time the wave brings to the market a great technology but the technology is not really able to gain a foothold. Either the niche was not large, the technology was difficult to produce or the technology was not scalable. In any case the wave fails quickly before the technology and products can take hold and build a defensible market share. One of the key issues to consider about technology is whether the economics are favorable; will it scale so that the products can be put into mass production. If the answer is no then the market is very limited and the wave will dissipate quickly.

Another problem that will cause waves to fail quickly is if the wave does not create or stimulate any early adopters. In this case, the wave may start but there is no customer traction and the wave will collapse because of lack of demand. One always needs to be mindful of the need to not only build a product but to build a product that someone cares enough about to actually spend money on. No matter how large a technology wave you think you're riding, or how good you think your product is, you have to make your product economically viable in order to get traction. In the mid-eighties I was consulting with a number of companies that were introducing new products. Invariably the companies would trot out a product slide that would estimate the market and the proponents of the technology would say that given the market size they would/could be very successful if they just got 1% of the market because of its large size. The problem is that not everyone can create a product that gets a 1% share and if the market is really that big the number of alternatives will increase dramatically and companies holding small market share will lose out. You must simply make a product that can garner market share over an extended period of time.

This brings us to the key problem with a wave of technology. To be really successful you must develop a technology and its resultant products have to get you to a high volume consumer-related product. You must achieve a mechanism where you have economy of scale otherwise you are at the mercy of the emergence of a large well-funded player who is able to come into the market and dominate the market. Simply put, a well-funded player like Amazon can afford to run with low or no profits for years until they gain dominant market share and only then begin to try to make money. If you are their competitor you must figure out a position or strategy that does not allow the juggernaut to crush you,

PROBLEMS WITH WAVES

If you are dealing with a small wave or are trying to exploit a niche market, you must figure out how to expand the market and achieve a critical foothold beyond just creating a small wave or niche product.

Chapter 17
Decay of Waves

Waves can decay/fail for a number of reasons. Two of the most extreme reasons are: the wave was never very large and thus it was just a small niche technology/product or the wave was too large. If it's too large it attracts so much attention and money that it compresses the product cycle. At that point you have a giant wave and when it crashes it creates a huge churn and consolidation of products and companies that can survive in that environment.

If the wave is small the decay of the wave is really pretty benign (as long as you are not actively involved in the industry generated by the wave). The wave never generates a large number of high value companies and thus when the wave decays and starts to churn up onto the beach there is not really a big economic effect unless you are an employee of one of those companies that lose out. In many cases this type of small wave has generated a number of small companies and there will usually be a market niche that remains and a number of companies will be able to succeed (i.e., make a small but reasonable living) even though the market never gets very large. As examples of changes in market niches consider the resurgence of a small vinyl record business. Rainbo Records is reported to have pressed nearly 7 million records last year. And, after opening in 2011, Quality Record Pressings is producing about a million vinyl records a year.

While the number pales in comparison to digital downloads of music, it is nevertheless significant.

It is the large steep big wave that can cause a lot of problems when they collapse. In this case lots of people and funds have a vested financial stake in the process and they will not go gently into the night. Economic wars essentially break out with winner take all strategies in place. This is because when the face of the wave forms after the initial break the wave rises up and the products begin to proliferate from the development labs. As the wave ages and moves to shore, the wave begins to become dangerous and the curl starts to form.

The shakeout begins as the curl forms. In the ocean, the top of the wave will begin to fall over the lower, slower-moving portion of the wave, forming a curl. In the business world, this is the beginning of the end. Eventually, the curl will collapse and the technology will start to be a commodity. Big money begets a large and consolidating industry. As the curl starts to override itself (and form a pipe), it is easy to get crushed. This is where you must decide whether to ride into the beach and try not to get crushed or to go back over the top of the wave and into the swells to try to ride another wave. There are different types of technologists who ride big technology waves.

White water finally forms as the wave breaks up and moves into shore. This is when the technology starts to be a commodity and only the strong companies with large production runs can survive. I do not find production interesting so I suggest that you find a good textbook on business management practices if you want to know how to run a commodity product company.

Fortunately, in the technology world there are usually other waves forming and you can find another big wave to ride.

This is really what you want to happen. You know that a wave will collapse and decay. So if you are adventurous you need to seek out big waves that have high stakes and see how far you can get your product and company before the wave decays and dies.

Chapter 18
DEATH OF A WAVE

Waves of technology eventually die. The wave may have decayed to the point that it is no longer viable. Still some companies and products can carry on for a time. But eventually the wave dies and the market will almost completely disappear. This can happen in a number of ways and take different amounts of time.

If you have a well-established wave that has generated a lot of product and corporate revenue, you can expect that it will be difficult for a new product or technology to displace the existing product. People are creatures of habit and they want to stay with the familiar. Thus, once a product is established you can milk it even if the technology changes. Look at the cutover from regular analog television to digital HD television. The changeover happened a couple of years ago. But, even after years of warnings and the "final" cutover day arriving, there were still a large number of channels broadcasting analog signals for at least another six months. When people's television sets finally went blank then the public finally paid attention to the switchover. And, this was really a simple changeover of a well understood technology. If you have a more complex technology it can take a much longer time. Unfortunately for the TV people many of them failed to recognize that the rapid sales growth in the television market was really just the replacement of existing sets. So they set their production schedules based on that replacement uptick and when most sets were replaced their sales took a nose dive with substantial economic consequences.

If there is an entrenched group of products in the market, the last thing the companies who control the products want is for the products to become obsolete! Thus, the manufacturers will come up with all types of variations to keep their products in the game for as long as possible. And, you cannot believe how long they can extend the life of a product. Just look at how long Sony's PlayStation and Microsoft's Xbox have been dueling with each other for supremacy in the video game market. A market that the Gartner Group predicts will grow to over a $110 billion in a few years. Buried in those numbers is a notable sub-trend as Gartner says mobile or casual games played on smart phones and tablets will rise to 20% of the overall market in that same time frame. That's a projected increase of 8% in two years and may harbor a trend that game playing habits may be moving in a different direction because of mobile devices. That's why Microsoft in particular is loading up the Xbox with added features for streaming movies, making calls, internet, TV, etc. essentially trying to make it a "smart box" for more than just gamers.

When a product dies it can die fast. This is analogous to a wave crashing upon a beach. The wave moves towards the shore with its structure pretty much intact until it curls over into a random mess of foamy water that is moving at a completely different speed. If the wave is a big wave and has been running for a substantial amount of time the wave will just crash with a massive amount of energy that can crush anything caught in its path. In many cases the wave just crashes by rolling over onto itself and rapidly dissipating its energy.

This is similar to the product business. When a product is finally obsolete, the buyers just disappear. Just look at the transition from VHS video tape to DVD. When the DVD market got going, video tape just died as a consumer product.

The same is true of the shift to smart phones and tablets. Right now it doesn't look good for the personal computer business. It may be on the ropes having been done in by the tablet. But there are lots of products and industries that go through these types of cycles. In some cases, industries even manage to reinvent themselves.

Consider a low technology business, beer. Well, actually, the beer business may nicely illustrate a significant technology issue. In the old days (mainly prior to the 60s) beer was generally a local product. There were some large national brands, but there were lots of local breweries. Part of the reason for this was that transportation of beer was difficult and costly because of the transportation infrastructure of the time. But the advent of two technologies (cost effective refrigerated transport and the upgrading of the transportation infrastructure with the interstate highway system) changed the equation. With these two developments, big breweries could consolidate the market and local beer brands began to disappear from the market. The effect of the consolidation was that beer in a majority of cases lost its uniqueness as the big brands homogenized the market and brought out very generic tasting beers. In Europe the situation was different in that many towns and villages prided themselves for their local, unique brews. But, in America, the big brewers could move beer wherever they needed and small breweries disappeared.

The tide began to change as brewing equipment became readily available and many small breweries and brew pubs started to appear in the 2000 time frame. Some even sooner like the Boston Beer Company, brewers of Sam Adams beer, which started in 1984. The result is that now there is a thriving business of small firms and people brewing a wide variety of beer with different characteristics.

In fact, the equipment that started this revolution originated with home brewers who moved into small venues and then into small brew pubs. This is a case of the reinvention of a product strategy. We start with local brewers. They get consolidated and the beer starts to taste the same. Equipment becomes cheap enough to create small batch breweries. The result is that unique styles and characteristics of beer return and are highly localized (in some cases available only in a specific neighborhood).

The small brewery business of the first half of the century died. The big breweries became dominant, but allowed a niche for small local breweries to reemerge. There is no sign at the moment that the small brewers will be able to take control of the market, but they have carved out a solid large niche (right now it's estimated that craft beer accounts for 7% of the total volume of beer sales in the US) with their uniqueness.

CHAPTER 19
VARIATIONS OF A WAVE

We've considered both the decay and death of a wave. But, we could try to alter the situation by developing variations of waves that are decaying and dying. If we see a wave start to decay and we are in the product space we need to act. We need to vary our approach.

We've talked about how waves can differ. They can have different durations as well as different heights and slopes. Additionally, they can move towards shore at different rates. This is an important parameter of a wave, the rate at which waves get launched at the shore.

If you look at technology waves, it is not enough to look at the height and slope; you must look at the rate at which the technology launches. Launch rate may be the biggest and most important variation. The best form of a launch is a hockey stick pattern. A short horizontal component followed by a rapid vertical ascent as was illustrated in **Figure 1**.

Simply look at the growth rate of Facebook and Twitter. The growth rate as the products launch is just phenomenal and the results are that in a short period of time (a few years) the user base grows to numbers in the hundreds of millions.

A technology that launches lots of different types of technology waves has real promise for opportunity. When technology provides a lot of variation and changes at a rapid rate, you have the opportunity to catch a wave even if you did not catch the previous wave. More importantly you do not have to wait long for the next wave to come by to hop aboard.

However, even more important, waves that are coming at a rapid rate generally have two interesting properties.

One is that the technology field is actually understood so that you do not have to do pioneering work to establish your product or position out on the bleeding edge to try to convince people of the worth of the technology. The people who came before you have already done that heavy lifting.

Secondly, you are usually working on a technology or a product that people are generally familiar with. And you probably know how to produce the basic product. Your job is to come up with a variation of the product that gets you a unique and defendable position. This is really a lot easier than dealing with this problem from scratch.

Further, because the rate of innovation sets up the timing of the waves, you have the possibility of being involved in a set of products where you might be able to establish the rate of change by controlling the changes you put into your products. Thereby, you might have some ability to control the product introduction cycle.

This is true in a lot of industries in the computer and phone markets. In many ways the product innovation and introduction cycle becomes quite predictable. Moore's Law pretty much dictates major advances in semiconductor circuits on a period of about 18 months. Apple and the smart phone manufacturers seem to favor a shorter cycle. Electronic game developers need something new each year for the Christmas season. It becomes somewhat predictable how wave cycles can be set up and understood.

Chapter 20
REINVENTION OF WAVES

The easiest thing to do if you are trying to catch a wave or develop a product is to set out to reinvent a successful product. Lots of successful products incorporated aspects of other successful concepts or products. Simply look at the idea of adding a camera capability to a smart phone. This was an easy idea. Once you make this decision you then face a whole series of tradeoffs concerning cost and quality but the basic idea is simple and easy. And, it has unfortunate consequences for the camera industry. which is then thrust into the problem of reinventing itself or dying.

In a large number of cases, owners of successful products want to keep the product going in the same direction for a long period of time. This is to their benefit. They do not want to retool the factory. They can keep or modify their sales literature. It is easy to maintain the status quo and slightly modify the system they are selling. In a lot of cases this is enough to keep you in a solid driver's seat with your product especially if you are at the top of the pile.

But, you need to constantly worry about the disruptive technology wave that can probably replace your product or alter the product space.

This happens very frequently. It may be a new product that combines the features of two previous products with updated technology. It might use technology in a completely new way. Or, it may be using technology in a straight-forward way but changing the form of the technology by making it portable. For example, tablets essentially use all of the above techniques. They combine a large number of functions. They change the form of the technology by not only adding functions but by extending the touch screen technology and getting rid of the keyboard. And, they change the portability equation and the types of software that run on the system.

In doing so they have essentially created a new industry that can actually destroy the personal computer industry. The tablet is a classic reinvention scenario and it is these types of disruptions that cause new waves of technology to exist and provide for opportunity.

Concluding Remarks on Waves

Trends can create many different waves. As time goes by many waves can spin out of a single trend as people learn to exploit the technology formed by the trend. In the case of the growth of digital electronics formed by the microprocessor, the technology moves from enabling just processor enhancements to providing for more advanced communication technologies. Coupled with a change from circuit switched systems to packet based systems, the internet is formed and an entire new set of end user applications eventually form. Waves of technology advancement can then ensue over a period of years.

Spotting a trend in hindsight is easy. Spotting it as it forms is tough! The growth of the number of connected hosts on the internet enabled forward thinkers the ability to create waves of productivity improvement delivered over the internet. In a very short period of time this constant arrival of technology caused waves of fundamental changes in the way we use technology.

You do not need to spot a lot of trends. If you spot—either as an inventor or as an investor—one of the major trends like social media on top of the internet, you can make a major change in your life and your family's economic status. Spotting a big wave can be a defining life experience.

My advice is to spot the wave and then ride the wave. But you do not have to try to ride every wave that you think is forming!

PART 3
WINDOWS

Introduction to Windows

In parts one and two we looked at trends and waves.

In this section we're going to take a look at how to catch a wave that has formed out of a trend. We will also discuss when the wave becomes visible and how we might profit from our observations. First, we determine if a window of opportunity has opened up and whether we are in a position to exploit the window.

Windows will provide the basic opportunity by which we can capitalize on trends that formed into a wave. Both trends and waves are tricky concepts to capitalize on. A trend may take years to develop before it can form a set of waves that possibly can become products. Waves may form but be of such limited duration that they cannot be exploited. What we want to find is a trend that develops a basic technology that lasts a long time. This gives us a chance to take our observations and turn them into some money.

Forget the pundits. If you listen to the radio and television, or read the written word, the economy is in bad shape. We must be on the verge of disaster. Au contraire, forget those pundits. Ignore the politicians. There is tons of money in the world. The world is awash in money. Money begets money and the big boys are financing all kinds of new inventions. They have a fiduciary responsibility to invest the money entrusted to them and to try to multiply it. Sure we will probably have glitches when the system clogs up and may even grind to a halt, but if you think that the investment bankers are going to stop trying to make money (really big money) you've got it all wrong. They have families, fast cars, big houses and yachts. They will reinvent whatever they need to reinvent to keep that endless drive to collect more money and toys. Just look at the money made by the people who created our recent mortgage crisis. You could call it the opportunity crisis. They made money creating the crisis and they found a new window of opportunity—unwinding the crisis. An instant trend was created and a big wave formed. The window immediately existed. How do we unwind the mess? Instead of throwing the evil doers in jail and throwing away the key, we reward the bad guys. The only guys who could sort out the mess were the guys who created it. They got to profit by creation and profit again by unwinding. And, all the while they were using money from the taxpayers because we let them get "too big to fail." But, could you profit at that exact moment?

Once the Fed decides that the big banks are too big to fail and fills them up with money, you too have a window of opportunity. Your window is more limited because you only have the ability to buy publically traded stocks of banks and mortgage companies, but you have a window of opportunity. The only complaint you may have is that your window is more limited than theirs. They are flat out in a position to make big money, but so are you relative to your income.

Their window creates your window!

The question is, do you have the ability to act and respond? This is tricky because at the time the window opens up, the world looks like it is going to explode and fracture into a million little pieces of space dust. We were on the verge of a full-out depression and the only way out was for the big boys to sort it out. And, if you think that they were not going to profit on that window of opportunity, you should think again.

This illustrates a key point about windows of opportunity. Generally they occur during times of unbelievable disruption during which you may be fighting for your very life and livelihood and you may not be in a position to capitalize on the window. Hindsight will always be 20/20 and you will miss many windows.

No matter, there are always windows opening up and you probably need to catch only one to really position yourself for a great life. On the other hand, professional investors and investment bankers need to catch a lot of such windows to stay on top of their game and bring in the big windfalls necessary to support their lifestyle.

One of the advantages the little guy has is that if you spot a window of opportunity you can take advantage of it. Let's go back to the moment of the financial meltdown in late 2008. At one point, as the banks were melting down, any company with any mortgage or financial exposure got hammered. Among other firms, Lehman Brothers investment bankers went out of business. Look at General Electric (GE). GE is a great venerable company with a wide variety of products. Because of its financial arm, it came close to being annihilated during the financial crisis. If you had the foresight to look past the crisis and bet the farm on GE you, (just like the big boys), could have capitalized on a once in a lifetime opportunity. The window that you needed to spot was the window that was created by the panic that GE would go under. It might have. However, when the Wall Street boys in the federal government started to bail out their too big to fail buddies on Wall Street everyone, even the people on Main Street, could got their shot at the big time window. But, the Main Street people would not find their opportunity financed by the Feds. To reassure nervous shareholders GE got Warren Buffet to invest in the company during the height of the crisis in October 2008. Three years later it was estimated that Buffet made over $1.2 billion on his initial $3 billion investment.[3]

The window was there; I just never said that everyone gets an equal shot at the window!

---

3. Shira Ovide. *Warren Buffett Made at Least $1.2 Billion from GE.* Blogs.WSJ.com/deals. 13 September 2011

Chapter 21
THE CONCEPT OF WINDOWS

Windows are really easy to understand.

The basis of a window is that at a given time a technology will provide groups of people the opportunity to enter a portion of the market with a reasonable probability of establishing a viable beachhead in the market. Sometimes people refer to this as the "window of opportunity". This is really the time that you have to make your stand in the market. Otherwise the probability that you establish a position becomes very slim. Slim, not impossible as you can always enter a market, but whether you can do it cost effectively or profitably is a completely different question. With some technologies you may get multiple windows of opportunity (e.g., cell phones as they transition towards Smartphones) and with other technologies you may effectively only get one shot (operating systems such as Windows).

There may be players in the market before the window opens up and they could be the proverbial pioneers that end up being doomed. They got to the market before the potential consumers of the product—the buyers—realized the importance of the product and actually began to buy it.

There may be players entering the market as it starts to

consolidate in which case the new player must bring something extremely important to the market or inertia will stop them from being able to enter the market and establish a viable position.

If the market gets big enough and complex enough, the dominant players may not be able to cover the entire product space with quality products. Windows may open up for niche products that push the product space by concentrating on issues such as performance, low cost, less weight or other unique features.

The basis of the window concept is whether you can realize that a technology has just formed or is about to form a wave. That allows you to enter the market with a shot at making product sales. At the same time it might help you avoid being relegated to the scrap heap when the consolidation occurs.

If a market is big enough, you can bet that the second the market starts to take off there will be a lot of products brought to market. And, you will have to run fast. Yet, it is possible to enter the market at a second window if such a thing happens. In the early eighties, I was working on ideas associated with local area network (LAN) based systems. I used to laugh about the market size. At one point there were literally hundreds of companies trying to enter the market and establish their positions with a variety of technologies. Usually when I talked to the presidents of these competitors they would tell me that the market was so huge that if they could just capture one percent they would be satisfied. But, they did not take into account either the level of competition they were facing or the actual size of the market in terms of number of units that would and could be sold. There were a large number of competing technologies as well groups of companies piling into each of the technologies. In this case the window was open if you could get your technology established and running in front of the other technologies. But the window quickly started to close as the types of LANs (ARCNET, Token Ring, Ethernet) being considered dwindled and the companies with dominant positions started to consolidate the industry.

Even companies like IBM, which tried to enter the market late (my opinion) with their token ring technology, got overrun by the

Ethernet gang and the Ethernet gang saw a corporate shake out and only a few vendors survived.

The window of opportunity is like a real window. It can be opened; but, it can be slammed shut and you will bust your fingers if you have your hand in the open window when it comes down. Importantly, just like a real window, there may be a later opportunity for the window to be reopened.

Chapter 22
THE IMPORTANCE OF TIMING THE WINDOW

Windows are very important. They define the point or the opportunity when we can bring a product to the market. No matter what the trend or the wave if the window exists there is a real possibility that we have an opportunity.

The point at which this opportunity opens up is the start of the window. The far side of the window is when the opportunity starts to fade (due to competition, lack of buyer interest, etc.). When that point occurs the window will close with a bang.

Windows of opportunity are not difficult to spot—in hindsight; they are always around. However, whether you can actually spot and act on one is an entirely different matter. You simply might not be at the right place at the right time.

I was once in a meeting with the representative of a major aircraft manufacturer and we were discussing various technologies. Usually in such meetings we might have a positive result and a mutual interest in certain categories of technology. But more often than not, the result is that the big vendor views the technology we are trying to sell as too late to the market to be really interesting. In this case after discussing the technology the potential customer thought we were too early in the technology cycle to be interesting.

That is the problem with windows of opportunity. To succeed the timing must be right. You cannot be too early or too late. Either case is the kiss of death.

The window defines your opportunity. You must hit it at exactly the right time or it will close (too late). If you try to exploit it before the market is there no one will be interested (too early).

Windows define your opportunity and you have a limited time to spot one and try to exploit it with your product.

Chapter 23
Window Types

There are several types of windows, but I like to think of them in terms of their maturity and size. I believe there are lots of opportunities in the world to catch a wave by observing the waves maturity and size. I do not want a large and mature wave if I could catch a wave that was just starting.

On every wave there is a window of opportunity. The question is can you capitalize on the window when it is open. Consider the tablet market. When Apple introduced the first successful tablet, the iPad, a big wave was started and windows opened up. Similarly, when Apple introduced the iPhone a big wave was started and windows opened up.

Let's go back over that for a minute. When I talk about the opportunity presented by these product introductions I noted that windows (plural) opened up. In both cases, several windows opened. Apps are a classic example of a set of windows opening up. In both cases, an entire field—that of Apps—became available for entrepreneurs to innovate and develop products. You could develop a wide range of products from games to news to financial applications. The sky and your imagination were really your only limits.

In the case of Apps, at the introduction of the iPhone the field was essentially wide open. How well you did was determined by how fast you could innovate and position your product. By the time the iPad arrived your choices were limited because everyone who had an iPhone App would want to port it to this new platform. You also had a number of new entrants who wanted to enter the market even if they didn't currently have an app. People understood there was a viable platform for application development and by the time the iPad was introduced they did not want to be left out.

The iPhone/iPad products make a good case for the importance of understanding different types of windows.

At the introduction of the iPhone, the market for Apps was immature (compared to today) and we were not sure of its size. By the time the iPad was introduced, we know that the market is large, but the unit revenue is probably small for each App so you have to have an App that is in high consumer demand. The result was that for a reasonable period of time the App market window was open to almost any skilled software developers, but began to close quickly as the market matured very rapidly.

In the case of the iPad, the window opened, but the number of software developers with entrenched positions was quite large and you had a large hurdle to jump over to get through the window successfully. Many of the easy/obvious/essential Apps had their basic components developed and for a developer to move its App over to the iPad would be much simpler than developing a completely new App.

Looking at the hardware side, namely the introduction of a tablet, the tablet wars were just beginning but developing a competing tablet would be difficult as the bar was already set for the consumer. They would measure your tablet against the iPad. Similarly, the iPhone had the same effect in the smartphone market.

As an example of the dynamics of windows consider that at the introduction of the iPhone, Blackberry had approximately 50% of the smartphone market. By mid-2013 (just six years) Blackberry had less than 4% of the smart phone market. With dwindling finances how could Blackberry jump into the tablet market. Even Samsung who had a solid smart phone market share would have trouble getting into the market successfully.

Even the Amazon Kindle would have trouble competing against the iPad as Amazon had positioned their device as a single purpose reading device. In Amazon's case they have the "luxury" of being a big player and continue to improve the Kindle (and destroy the MS/Barnes & Noble Nook in the process) with new versions like HDX that has higher resolution not only for reading but also for watching videos. Amazon is tying Kindle to its Amazon prime strategy where for a yearly fee you have access to their library of over 40,000 TV shows and movies. They also use it as a portal for their customers to buy other products. It's been reported that shoppers who own Amazon's tablets and e-readers buy 50% more often than those that don't. Now the direction for Kindle, the "formerly" single purpose machine, is clearer.

As the wave shifted and windows opened and closed there were opportunities for the really big guys in the hardware space. There were opportunities for App developers and there were cross platform development opportunities.

What all the jockeying around masks is that the Smartphone market and tablet market really began to develop and open a series of windows that are defining the mobility space and causing a completely new set of markets to develop. It's now BYOD (Bring Your Own Device) in many companies who handle mobile computing in the cloud.

For many new concepts a wide variety of windows open up. For an individual App there may be a small opening of short duration. For related groups of Apps there may be a series of windows opening based upon the developer's ability to evolve as the market changes. These windows may be of significant size and duration. Lastly, for hardware and platform developers there were/will be a set of hardware windows that open and close but they are of significant size.

Regardless of the size of the window, the markets and opportunity for Smartphones and tablets are opening and closing quickly because of the rapid push of new products by the manufacturers,

Other opportunities like integrating these technologies into enterprise scale systems will create significant opportunities for the foreseeable future.

Chapter 24
Definition of Windows

Windows are really easy to understand. The concept of a window of opportunity is also pretty simple.

The concept is easiest to visualize in physical terms. A good example is a double hung window that has an upper part and a moveable lower part. An easy way to open the window is to raise the lower part so you can let in air from the outside or exhaust air from the inside. But, depending on the type and age of the window, the raised part may have a tendency to want to close itself.

In some older double hung windows, the counterweights tend to fail and unless you are careful (and prop the window open with a dowel) you can experience a situation where the window suddenly slams shut. If you're unlucky and have your hands in the wrong place you can get your hands broken. This is the same problem you have if you try to go through a window of opportunity and it closes on you. But, in this case it is time and money that you lose.

Trends are precursors to waves. When a wave forms and people recognize it, companies and products can be developed. But, as a wave forms you need to be able to put yourself in position to take advantage of the wave. This is where the window of opportunity is important. You need to find a product or corporate window on the wave and jump through the window so that you can exploit the opportunity.

Windows may exist for a short period of time and they can close very quickly. You have to be nimble to jump through a window and if you are not nimble the economic consequences can be particularly serious. As a wave forms many people will want to invest to develop companies and/or products, but only a few will succeed. It is not enough to recognize that a wave has formed. It also requires that you can see a product or corporate window of opportunity. You must be able to get a product into position where you miss the time curse. Remember, the time curse is when you're either too early or too late to market.

If you're too early you may get there before people recognize that the market exists. If you are too late the market may be so mature that there is no entry point. Timing is very critical. You need to go through the window at the right time.

I once saw a case where a startup company was about to introduce a new product. However, days before they could get to their announcement, *Business Week* magazine announced that the IBM PC was already the market winner in the personal computer (office automation) product category. This was a classic case of being too late. The company had planned to introduce a complete office automation system based upon a proprietary office automation system set on top of a proprietary PC.

Even if they had introduced the product a month earlier they could not have stopped the declaration. However, once the declaration was in the public domain there was no need to introduce their product in its current form. The winner had already been declared and the best the company could do was say "me too, but I am proprietary". Maybe, they could even cite some advantages, but the winner was declared.

Being too early can also be a disaster. In many cases, if you look at product categories and study the history of product introductions, you will find that the first products were not always the most successful products. This happens all the time and it is very frustrating for product developers. The IBM PC was a follow-on and blew away the Apple II, the Commodore PET, and Radio Shack's TRS-80, just to name a few. In turn the IBM PC would also get payback from Dell and Compaq, causing IBM to eventually abandon the PC market. But then they struggled also as tablets emerged.

Another case study of windows opening and closing is the evolution of Netflix. It started as a shipper of DVDs and had to evolve to an on-demand content provider with proprietary programming. The DVD window basically closed and the on-demand window opened.

The key is to try and find the window and recognize that the window and its opportunities are shifting so it is like trying to jump through a window when it is opening and closing but moving through time.

People have a tendency to project that tomorrow will be like today. If there is something that you need to conceptualize it is that change is constant. I once dealt with an architect who had an interesting position. He felt that he could design the perfect office building. He worked hard and diligently at the process and was taken completely by surprise when the client immediately began to change the office configuration when the building was complete. In the few months that the design and build had taken, the firm for whom the office had been built had changed significantly and the original design no longer fit. You need to be flexible and understand change will occur.

Chapter 25
Visualization of Windows

In less complicated times, windows were difficult to spot because they were pretty closed; i.e., manufacturer's products were proprietary. The prevailing mindset was to develop products that were proprietary and to lock the customer into a particular set of products. This provided a framework where the window of opportunity was open for a short period of time when you could compete with the entrenched manufacturer.

New car introductions, where car manufactures tend to bunch their new car releases into a short time frame, are just one example of such windows. However, the introduction of an electric car, such as the Tesla, (being unique it has the luxury of creating its own time line), would not have to be introduced at the same time as car vendors introduce their normal product lines.

This is the case of what I call a simple window. The window is open and you either jump through it or you wait another year (if you are selling a regular model) because all the new cars generally come to market at the same time. This type of window can be visualized as opening and closing several times a year based upon the industry.

Similar windows can be seen for pricing strategies. Again, cars are a good example as the car companies tend to have sales and run special price offers at about the same time each year. They want to attract people into the showroom at specific times of the year.

Similarly, department stores and discount stores tend to employ strategies that revolve around timed product (and sale) introductions.

It is easy to visualize such windows as they are either open (and you are ready) or they are closed. If you are not on the same time line as other product suppliers when they introduce their products you either have some heavy lifting to do to get noticed, or you have something really unique (like Tesla) so that it gets noticed.

When visualizing a window that is opening you have a problem. If you are alone in trying to jump through the window you must have something that really distinguishes your product. Otherwise you probably need to wait until the window is open very wide.

Both strategies are prone to problems. You may not want to be the pioneer, but you do not want to be viewed as a follower. Visualizing a window requires that you not only see that it is opening up, but you need to figure out how long it will be open. You need to determine how fast and when it might open and how fast it might close and whether or not you can have your product ready within the estimated time frame.

Visualization of the window is often easier than spotting a wave as there may be multiple windows opening on a wave as the wave forms and progresses towards the shore. At the same time each window that opens on a wave may open and close quickly and you might have to react fast before the window closes.

In some cases, there may be a series of windows with different properties that you can try and exploit. There are also cases where the window opens and the product space is seen as so lucrative that lots of large and aggressive companies have been waiting for that window to open. In fact the number of people trying to squeeze through that window is so large that sometimes it's almost impossible for you to figure out a strategy and get through that window. Look at the low-end computer printer market that's dominated by such heavyweights as HP, Epson and Cannon. Smaller competitors like

Brother have to scramble on features (lightweight, faster, wireless, paper size) and cost to try and compete.

Figure 7 illustrates the rapid initial growth of Google searches when search became a major Internet function. In this case Google essentially beat out other search engines (primarily Alta Vista, created by three researchers at the once mighty Digital Equipment Corporation or DEC) with the use of paid advertising and built the base necessary to dominate the search market. Currently, Google has over 65% of the search market in the US and dominates, almost like a fortress, the market for search.

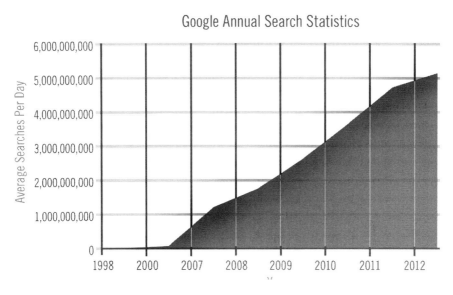

Figure 7 Illustration of the growth of internet search. (Source: Google)

Similarly, Facebook starting in 2004 with a small base of users has come to dominate social media. It has also had remarkable growth and along the way took Myspace out of the number one spot. In this case there are more specialized social media sites (LinkedIn and Yelp) that provide significant, but specialized competition to Facebook. But in terms of market value, Facebook is a serious leader in social media. This growth is illustrated in **Figure 8**.

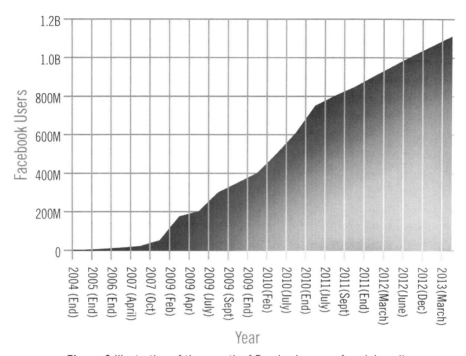

Figure 8 Illustration of the growth of Facebook users of social media
(Source: Facebook)

CHAPTER 26
PROBLEMS WITH WINDOWS

When we talked about trends and waves in previous parts of this book, we were talking about potentially long time frames. Trends can run for a substantial time and multiple waves can form. Windows of opportunity generally have a short time frame. It's not very often that a window of opportunity will be open for a long time. Multiple windows may form on a wave but each one may close quickly.

There are two major problems with windows. One is they can be easy to recognize and the other is that they can attract lots of money trying to exploit the opportunity. Prior to the Internet becoming a daily necessity of life for everything from social relationships to shopping, time frames were much longer. A long time might be measured in years and short time periods might be measured in months. Now a long time seems to be a year and short time seems to be measured in days.

When a product category starts to take off and is well defined there are a lot of people who can see the window. They understand that the window is open. In many cases, during the development of a technology when the product window opens up, you see a lot of possible product entries. One group who sees the possibilities is venture capitalists. The question I hear all the time from venture capitalists is "how many competitors are there and how big is the market space?" If there are competitors and the market space is well defined, then the VCs have a tendency to jump into the fray and build a company or develop a product.

Most people think about a window after it's defined (and a market is a defining characteristic). When it's defined they want to jump into the market. This happens all of the time. But if you can describe the market that means that the window has been open for a period of time and the window will close earlier than you can imagine. This does not mean that a similar window will not open shortly thereafter, but when you can describe the market you are closer to the end than you are to the beginning. This causes a conundrum—VCs are supposed to be looking at emerging products and companies and their big claim to fame is the development of new and interesting companies. In reality they tend to be followers. Just consider such companies as Apple, Cisco and Microsoft. VC money did not flow into those companies until they were fairly well developed.

When VC money starts to flow into an area, the window has been open for a time and you need to be really careful.

As I write this, the US is still trying to emerge from a major recession and there is high unemployment. However, in my opinion the world is awash with money. How can this be? I accept and believe in the premise that there is constant change in the world on all fronts. Jobs and their requirements are always changing. Some changes are not correlated. The level of unemployment, the strength of the economy and the amount of capital in the world are not necessarily related. Many product categories and companies start during a recession or at a time of high unemployment. A recession actually encourages start up activities as investors (if they have money) have to look harder to find good investments. They also may be willing to take larger risks. But, they may be more demanding on the terms they want to extract from corporate founders in return for the startup money.

In a situation where investment opportunities equal higher risk, then investors will demand a higher return on their money because of that risk.

When we are faced with a bad economic environment, products or companies that promise increased productivity and lower labor costs will tend to be favored in the race to get products through the window of opportunity. If you look at companies formed during recessionary times you see companies with productivity themes such as Oracle (database tools to enable better information processing) as well as Microsoft (office automation productivity tools).

There are many examples of companies that are formed in bad times. Companies like Square or Flint, which are poised to disrupt the financial payments industry, are just one example of companies that arise during an economic downturn.

Not all segments of the economy may have money when a window of opportunity opens up. There are many times that a window will selectively open to some investors and close to others. If you're an investor and have overcommitted to investments in an earlier window that is not working out very well you will end up having to support your previous investments and not be able to take advantage of currently open windows.

Trying to jump through an open window is a very tricky proposition. It can be costly if you get caught in a war between other companies or start a war yourself. You have to be very selective about when or where you try to jump through a window. The biggest problem is that you do not know how big the market is so you don't know if you have a real opportunity (compared to the risk). Unless you have good research (or intuition for the market and product direction) you have no idea how many other companies will be trying to jump through that window.

Chapter 27
Decay of Windows

Once a window is open it will start to decay. This is usually a precursor to the window closing. But it may take a long time for the window to close and it may open, decay, reopen, decay, etc. for a long time before it eventually decays and closes. The window will close up. It is just a matter of how fast.

What happens is that the opening of a window sets off a race of products that causes the window to start to close. But, if the window is large enough there is then a process of change that could cause the window to open wider. Or, the window may nearly close and then it may stay open just enough for one or two companies to continue to innovate and keep a steady stream of products and innovation coming. For the companies that are through the window and are producing successful products and revenue, they can probably stay in the game for a long time. Companies like Apple and Microsoft will try to keep the PC market going as long as it stays a cash cow and allows them to expand into alternative, growing markets like search and games for MS and phones and tablets for Apple.

We do not know how long the window of opportunity will last. We have no idea how wide it will be and how long it will stay open. If the window opens and lots of companies jump in with products, we may have a situation where for a long time a number of companies stay alive, but just barely, while the market starts to settle out.

If the market is huge, or perceived to be huge, money will flow into a subset of companies and the market may collapse quickly. There are many cases of innovation where big money gets behind three or four companies at about the same time in the same product space and the war begins. The reason for this is that the investors in each company thought their product was unique and a number of smart people came to this conclusion at the same time. There really is a lot of money chasing good ideas. Technology extensions are relatively easy to come by and it is very common for similar ideas to suddenly develop at about the same time. Investors tend to follow the crowd. It's easy to explain an idea that is an extension of a current idea, and then investors pile into the product.

This can set up a battle between big well-funded companies and keep the window of opportunity open and decaying for a long period of time.

Even more interesting are situations when a company misses an opportunity, sees how big the opportunity is, and then tries to enter the market. One of the best examples of this was Microsoft who missed the initial Internet wave and has been trying for years to catch up. Although they will not say this, one can make an argument that Google's entry into the office automation market with free software in the cloud may have been a reaction to the attempted incursion into search with Bing by Microsoft.

The effect of these efforts is that the window changes and evolves, starts to decay, and then periodically opens back up as companies pour more money into the market as they try to encroach on each other's markets. In most cases no amount of money will allow a company to catch up.

A lot of big companies do not report separate divisional revenues. So they can hide a lot of sins as they try to start a product in a window that's just barely open and almost fully decayed to the point where it will stay closed.

You can only get a feel for this as you watch the numbers that illustrate the market share of the warring companies.

Institutional investors want the companies they invest in to grow and one way to grow is to develop a product that takes market share from another company that is in a hot market segment. But the hotter the segment, the more difficult it will be to knock off the top competitor.

There are examples of companies that reopen the window and knock off the top competitor. These examples keep entrepreneurs upbeat about their chances, but their chances are really slim.

The prime example of this is the fast food franchise industry. If you won a fast food franchise you probably want to locate near other busy franchises. Traffic is already established and you want to be nearby so you can steal sales if people's desires change.

It's no different with high tech companies. They want to jump through the window. It doesn't matter if it appears that it's closing because they believe that people understand the basic technology and may gravitate to their product if the window stays open longer than most people think.

Chapter 28
DEATH OF A WINDOW

At some point, the window of opportunity must close. There are several reasons for this to occur: the market may not develop, the market may not be very big or if the market is big it will be dominated by a few big players. If a window is starting to die, you must get out. You need to make sure that you understand that it is best to get out when you have the chance. If the window starts to close or people perceive it closing you will not be able to get out.

Just because a lot of smart people believe that a market window for a specific type of product will appear, it doesn't mean it's going to happen. The smart people are not always right. Most of the time, only in hindsight, do you find the answer. There are a lot of reasons why some things fail and others never materialize. Sometimes the window doesn't open for very long. This might mean the technology never morphs into a product (solar technology has been touted for years but has only recently begun to catch on in any meaningful way). No matter how smart the developers are or how credible the plans, there are just some windows that never open very wide or for very long. There is a very apt joke from the product development business that goes something like this—it is a good thing that more products are conceptualized than prototyped, prototyped than put on sale, and fail before they succeed or we would never have a job.

Another problem that can kill a window is if the market is just not there. You can study, you can analyze, you can predict, you can perform all kinds of analysis and you can know in your heart that you are the thought leader, but the market will let you know if you are right or wrong. And if you are wrong the window slams shut. It literally dies. This is a frequent problem in the product development business. Even if you are right the market may not develop the way you think it would or should have. You are not only competing against other competitors but you must compete with other windows. For every product there may be alternative or parallel technology solutions. It's not always clear you can defeat the other products in your window—never mind products from another window. A simple example of this is to ask why PDAs (Personal Digital Assistants) failed. Well, compared to the Smartphone market, the market was just too small. Add to that the PDA was not very easy to use and we end up with a few competitors, and a relatively small market. With the advent of the Smartphone, these products end up as museum curiosities.

Let's look at another scenario. If the market is big, a huge amount of money will flow into products as the market develops. This creates some tendencies.

First, the market becomes quite crowded. Second, it becomes difficult for customers to sort out their needs against the competing products.

Third, if the market continues to grow then more money will be attracted. Lastly, as the market ramps up, the window will close unless a couple of large well- financed companies made it through the window. If that's the case they will get locked into a deadly struggle, constantly trying to outdo each other in terms of product capabilities, cost reduction and product performance. The window has died. However, for a few companies that have an established position the opportunity continues. If the market becomes extremely large and new customer requirements take hold, then other related windows may open up and the market will be a viable opportunity. However, at that point the market will look different.

A classic example is the dominance and success of Cisco in the network router business. Their success caused rapid adoption of technologies that created a need for very high performance routers. That opened an adjacent window that spawned some new opportunities and companies that provide network-based services. This led to not only the growth of other router companies (Juniper Networks) but also managed services and data centers and now the development of cloud based computing solutions.

Chapter 29
Window Variations

I began this part of the book by writing about windows as if there is one type and that it opens and closes in a systematic manner. In fact, the concept is a lot messier than that simple analysis.

Windows do open and close. Companies may try to get through a window of opportunity but decide to move to a different strategy, product or window due to a combination of factors. These could include the level of competition, the technology fit, new emerging technologies, lack of funds or other business considerations. But in recent years, due to the size of markets, windows have become very complex.

In some cases, the opening of a window for a large well-financed company will create a sizable number of adjacent, lucrative windows for smaller competitors. In a sense, it is like opening a double hung window in which the window is covered with lots of little panes that are within one frame of the sliding window.

Consider the introduction of the iPhone. When it was introduced, lots of new, albeit smaller, windows began to open up.

You could supply cases to protect the hardware and accessories.

You could supply APPs.

You could create entirely new products for a marketplace that had not been conceived before. For example, the use of a Smartphone as a point of sale terminal or a credit card processing device. Companies like Square, Intuit and others have jumped into that space in a big way.

If a window is large enough and developing fast enough, there are a lot of opportunities. The rapid sales of the base product will generate follow-on products, variations of the basic product and lots of opportunities for innovation in accessories and enhancements of the core product. The manufacturer will not be able to capitalize on all the potential products. They have their own problems coping with the growth of the base product and you will have a lot of opportunity to innovate in adjacent, similar or related windows of opportunity.

When people see the success of the base product, you can expect that competition will develop and it will develop quickly. People like to follow success and when you get a product that is widely accepted, people begin to think of lots of personal touches and add-ons they can make for the product. Thus a wide variety of windows of opportunity in different sizes, technologies and variations open up for exploitation.

In fact, if a manufacturer senses the success of their product but is limited with some of its resources, they will often assist developers to open other windows. They want to get as much traction and keep as many variations of windows open as possible. No manufacturer can control every aspect of the product environment or employ all the smart people. Many of them recognize this and try to pump up the market with outside help.

Chapter 30
Reinvention of Windows

The big question is what happens when you miss the window of opportunity?

Generally, you will see a lot of opportunity after it has passed you by and there is no entry point. This is a common problem, but it is not an insurmountable problem.

Many times, the first company through a window ends up with a bunch of arrows sticking out of their back. How many times does the first mover company get it right? The streets are littered with the bodies of first mover companies! If you are creating a new category of product, it is very difficult to get it right as you have very little experience with consumer reaction and purchasing habits. Your product may be too advanced or it may not catch anyone's fancy and it is dead on launch.

Think of how long it took the cell phone industry and the PDA industry to birth the smart phone. Once the industry was created, it looked like Blackberry had it locked up. Then along comes Apple and completely upsets the old apple cart. You just never know how things are going to end up. If you get to the top, you never know how long you're going to stay there.

Once you get through the window you need to understand that you have to run over your successful product even if you are on top of the world. If you do not reinvent and extend your product someone else will and that carries extreme consequences for your position.

An interesting strategy if you miss a window is to try and reinvent the window by changing the argument. As technology advances you can get an opportunity to pass through an open window or sometimes you can try to morph the window so that your product exhibits a unique set of characteristics and capabilities. You try to establish a new window that combines the features of products from several smaller windows.

Lots of times when a window closes, a similar window with a slight variation may open up. If you did not make it through the first window you will have a shot at the new window. Consider the introduction of tablet technology. The phone competitors to Apple were still trying to catch up when the tablet came out and then they had to run really fast.

But people who were through the Apple smart phone window had a leg up when the iPad came out. Still there was an opportunity for outsiders to break in. The suppliers were not necessarily small. Just look at the battle that Google, Facebook, Samsung and Apple are engaged in to try and dominate the mobile product space both as hardware and software companies.

The Smartphone and tablet really impacted a large set of software companies. But, the big market share loss was to Blackberry—their market share for Smartphones dropped from around 50% to about 5%. A devastating loss of momentum and it is not clear if they can get through another window of opportunity. Google has piled into the fray with their open Android system products, pushing the idea of phones and tablets with open interfaces so that anyone can develop for the product. Whether this reinvention works out is not clear. A big problem with Smartphones may be that the market may be beyond saturation. We will be able to see this issue clearer in hindsight as the market shakes out.

Reinvention is a tricky proposition in any environment but

when you have a high stakes, highly volatile market with big players, no player can afford to lose even one small advantage if they want to succeed.

Because windows can be reinvented, you must always be pressing forward and deciding where to position your next set of products both from a technology perspective and from a feature perspective.

Concluding Remarks About Windows

A wave can form and create multiple opportunities for the development of products and technology that can make money. When a wave has formed there are potentially many opportunities to profit. But the trick is to spot the opportunities early in the cycle before lots of other people begin to understand the technology and jump into the product fray. This provides us with the window of opportunity.

Windows of opportunity occur when technology variations form a wave that is really a product opportunity. If you can see a technology wave and learn to exploit variations of the technology you have a real possibility of seeing a window open up that you can jump through with a viable product.

Windows can both open and close. If you are late to the party that is forming around a technology window you can waste a lot of money because by the time you get there the decision is over and the winning products have already been determined.

You can also be too early to a window and get through it too soon before you can establish a serious position and monetize it. A classic example of this was Myspace which came long before (in Internet time) Facebook.

The ability to spot and assess the viability of a window of opportunity is a critical skill you need to develop if you intend to be a success in the product business.

PART 4
BUBBLES

Introduction to Bubbles

We are moving right along looking for trends that form waves and trying to see if a window of opportunity has opened up for us. Assuming that we are in the game, we face the Prophet/Doctor of Doom. Yes he exists and he is more than a comic book character. The stronger a trend, the bigger the wave (and the more spin-outs of trends and waves) and more windows of opportunity, the bigger the bubble will be. However, each window carries danger. The Prophet always reminds us that there is danger. The sky is falling. Doom is just around the corner. We are going to get killed by Wall Street. They will take our money. We are little and we are going to lose!

Here is the problem. The money guys on Wall Street are pretty smart. They went to the right schools. They have fancy computers. Everything they do is correct. At least that is the myth. But the Prophet knows better. Just remember there are two sides to every coin. Sometimes the big baddies get on different sides of a coin and they fight to the death. As a trend forms some of their fancy computers spot it. They start to invest and eventually everybody's computers spot the same trends and the herd mentality takes over. A big wave has formed. Success breeds imitation and eventually the space will become too crowded and a bubble will form! No one wants to be left out as they are generally measured in similar ways and they do not want their compensation damaged. At some point contrary points of view take over and there are big players on both sides of the coin. Only one side will be right in the end. No matter how long the end lasts.

Because there is so much money at stake it is only natural that if a big wave forms there will be so many people seeking a window of opportunity that someone has to lose money. The best example of this is the venture capital industry. Venture capitalists tend to be followers. They spot trends and when the window opens they jump in. Eventually the market becomes crowded. They find themselves unable to invest in new opportunities because they have to try to support their previous investments; with luck things eventually work out. This is classic where bubbles form on trends that seemingly have formed a set of big waves with lots of windows of opportunity. As one example just look at the disastrous IPO of Facebook to see a classic bubble that was the culmination of over investment in the social media space. Facebook launched at $38 a share in May of 2012. There were wild claims that the stock could easily climb to over $60 a share by the end of trading on the first day. At the end of that first day, Facebook closed at $38.23 a share. Facebook's stock lost over a quarter of its value by the end of May. The Wall Street Journal called the IPO a "fiasco." Now, of course, all is well as Facebook is trading in the $50 range and has been added to the S&P 500 index. But that first day and subsequent weeks caused a lot of people to lose a lot of money. They bet on the wrong bubble.

When something is too popular and it is the talk of the town—watch out for the bubbles!

Greed takes a long time to simmer. Fear is instantaneous and we can get burnt quickly. There is an old expression which fits here—don't try to catch a falling knife. The good Doctor of Doom notes that he is willing to try and jump through any window, catch any trend, ride any wave, but even he says that the key is to avoid being crushed when a bubble pops. When a bubble pops, there is no way to avoid getting trapped. If you can catch even small waves you are ok as long as you do not get caught by a bubble.

The advantage little guys have is that they can own relatively

BUBBLES

large positions that are not significant in the grand scheme of things. So they can avoid injury when the bubble pops by moving quickly to a liquidity strategy while the big boys try to sort out the mess.

Consider Warren Buffett. For years he has been the poster boy of quality investing. He invests as a business owner and is very successful. In fact, he is one of, if not the most successful investors in history. It's estimated that Buffett made $37 million a day in 2013. But he has to take big positions. If one of his investments goes south he needs to work it out. On the other hand if you had a portfolio that mirrored his and you saw the investment go south you could just (unceremoniously) jettison it. If he makes a mistake and gets tied into an investment that is involved with a bubble, he has to work it out. You on the other hand can just go off and find another investment.

For all of his success (and he is really successful) I know people who have done better on a percentage basis than Warren. And, their insight on their success is that the key to making money is getting out before the bubble bursts. You can take high risk positions on big waves if you can get out before the Prophet of Doom tries to take you out.

In big wave surfing the key is to have that Jet Ski driver nearby to pull you out of a disaster rather than letting yourself get pounded and crushed under the breaking big wave as it turns to churn and the bubbles form all around you.

Compared to investors such as Warren Buffet you as the small investor have the advantage of nimbleness.

Moving on we will examine bubbles and their properties in more detail.

Chapter 31
CONCEPT OF BUBBLES

Bubbles form when everyone thinks they have a sure thing; people start to believe that all they have to do is just enter the market and automatic investment riches will follow.

Bubbles are the aftermath of a sure thing where you can invest in products (develop products) without regard to economic reality. The best way to think about a bubble is to see it as the classic greater fool theory—no matter what you paid for a product, you can always find another fool and sell it for more than you bought it.

The process is like blowing a bubble with bubble gum. You keep pumping air into the bubble and eventually it gets thin enough and the bubble bursts. The economics are the same. More and more people come to the same financial conclusion and they continue to pump money into the system even as the economics may be changing. Suddenly the market drops, the economics burst and some number of people are holding the bag with nothing left for their money or efforts.

Bubbles can take a while to form and sometimes they form quickly. The housing bubble of the 90s-2000s formed over a long period of years. It took a combination of events to get things rolling including liar loans and cheap money. However, once the trend for housing prices to always increase became a valid truth that almost no one questioned, it was the beginning of the end. Then the market for housing collapsed. At the end, almost everyone was crying in their beer about being ripped off by the banks.

Markets collapse for a variety of reasons and you never know what trigger will pop the bubble. Consider the mini-bubble of the HDTV sales increase. For years the television industry was converting to HDTV. At the same time screens began to get larger, thinner and cheaper. The sales volume increase was really driven by a number of factors: the declining cost of better television sets and the need for replacement of analog television sets. When the conversion finally came, a large majority of televisions had been upgraded. In reality the market had been puffed up by two factors and one of them would not be valid going forward. So the industry had to make rather severe adjustments for declining volume particularly at the retail level. When almost everyone had converted, the market demand just dried up and the game was essentially over.

This happens a lot, people project the recent past directly onto the near future without understanding the fundamentals behind the causes of the trend and they get burned as the bubble bursts. The fact that you can blow a bubble that is 6 inches in diameter does not mean that you can take it to seven inches without it popping.

However, in a complex economic environment, where you are not the only player, you do not control or know what others are doing. The bubble can puff up quickly and pop rather quickly. In fact you might not even be aware of the other players until the bubble has popped. In the product business you can be surprised when an unknown competitor suddenly jumps out with a new product. A competitor can also surprise you if they decide to try and take market share by simply cutting prices to an unsustainable level to force consolidation of the market. The advent of price comparison software on mobile phones is also putting pressure on the retail industry and forcing companies out of business.

Chapter 32
IMPORTANCE OF BUBBLES

Bubbles are created when the opportunity is so large that a lot of players enter the market and things get out of hand. When lots of players start to build their business structure on just getting market share solely on the basis of price you will get a bubble in assets or investment and it can only end poorly. Whether it is tulips, hand-held calculators, house prices or internet pet food sales the end is coming. The problem is that we will never know until after the fact when the end has arrived. Until then the general business strategy is the "greater fool theory" which works really well as long as you are not the fool. This theory assumes that you can invest without regard for the price of an asset as long as you can find someone who will pay more for the asset than you did no matter how ridiculous the price you paid—the greater fool.

For everything to go right, you must avoid the bubbles. If you get caught in a bubble you entire financial health can be destroyed in seconds.

Yet bubbles are a naturally occurring problem with examples throughout history. It seems to be human nature for excess to be the normal state of affairs.

Bubbles are a natural ending for a string of events. If people are successful at pursuing a particular path, others tend to want to follow. If you can develop a pattern and some level of success, then others will follow. It is this act that can generate a feeling of complacency that allows large groups of followers to adopt a rosy outlook for future events. This rosy attitude will eventually lead to problems.

The reason bubbles are important is that you need to recognize and avoid participating in a bubble when it gets over extended.

Human nature makes a bubble very difficult to spot. Did you spot the housing bubble? It was very difficult to spot. In fact, you would not really spot it unless you knew about the "liar loans". It's hard to spot a bubble like the housing bubble without having some of the underlying facts.

If you looked at the housing market pre-collapse, you might conclude that the market was hot due to demand. But would you sell your house, rent for awhile until the housing market crashed, and then buy back another house at a much lower price? If you did you were a very rare breed and must have had a lot of insight into the housing market.

Demand for houses was very high. Construction costs were going up, but where was the top. If the regulators and banks were doing their job and if the people who were packaging up the mortgages were doing their due diligence, everything should have been ok.

There were a lot of articles going back at least five years before the housing bubble burst talking about the overextended housing market. How many people acted? As long as the bubble was expanding what was the problem? Things could have worked out but for the fact that the whole thing was a house of cards.

There were a few people who actually spotted this whole problem and a few of them acted. Michael Lewis' book—*The Big Short*—famously profiles those that did understand and act like Dr. Michael Burry of Scion Capital, Jamie Mai and Charlie Ledley, the founders of Cornwall Capital, and others who bet against the housing market and made millions, sometimes billions as a result. The majority of people and institutions did not act or understand.

At the end, the bubble burst. And, a lot of people got injured due to the poor performance, bad behavior or crooked behavior of the participants in the scheme to kite the housing market.

If you had known that large numbers of financial firms were engaged in fraud (my personal opinion) you could have escaped the fiasco, but unless you knew the extent of the liar loans and shoddy underwriting you would not have been able to escape. Many luminaries writing for financial journals talked about the housing bubble but almost everyone misunderstood its size and impact. Many had no idea about the CDOs (collateralized debt obligations) and other exotic financial instruments hiding beneath shady "no-doc or low-doc loans." (Warren Buffett has famously called some of these exotic financial mechanisms the equivalent of "financial weapons of mass destruction.")

As conventional wisdom takes wing and begins to be viewed as gospel, you need to make sure that you are on the lookout for a bubble as bubbles will inevitably burst!

If you are in the technology product business you must be on the lookout for bubbles forming and reposition your product constantly. If you are too successful you will suddenly find that you have a lot of new competitors. Your only choice is to try to change your product before the bubble bursts. Change is always difficult but product markets are difficult to milk and can rapidly decline in value as new products enter the market and change the market dynamics.

Chapter 33
TYPES OF BUBBLES

There are numerous types of bubbles. They can occur almost anywhere. But they all come to the same end. They pop.

Bubbles can be large or small. They can involve different types of assets and they can have different rates of growth, but they all end in disaster.

Consider the size of a bubble. If you want to measure a bubble you probably want to measure its size in currency. That is a good way to measure a bubble and to measure its financial impact. The housing and financial bubble of period 2007+ took down a couple of brokerage firms (Lehman Brothers and Bear Sterns), and took down a couple of banks and destroyed billions in housing equity as prices had to readjust to the new reality.

If the Fed did not intervene, it seemed possible the entire economy could have gone down. However, in my opinion, this was not a real issue. Just because something/someone goes bankrupt does not mean that the productive assets disappear permanently. Had the Fed let Goldman go under would the economy be much different than it is today? In my opinion, no. In the Fed's opinion, yes. But since I was not running the place we did not get a chance to find out.

In my view, one key type of bubble is the bubble that destroys equity positions. Essentially the worth of an enterprise disappears. If Goldman had gone under, a lot of equity would have disappeared, but not much more than that would have happened. The people whose equity disappeared would have been injured, but it would not have necessarily have brought the world down or put us in a different place than we are today.

Enron was a bubble. Actually, it was a relatively small bubble. Yes, I know one man's recession is another man's depression and I feel sorry for the unfortunates fleeced by Enron. I also feel sorry for the unfortunates fleeced by WorldCom and Bernie Madoff.

There are always bubbles. They come and they go. Size is a critical issue as some can be very large.

Look at all of the net worth that disappeared due to the Internet/y2k bubble. This bubble was bigger than Enron and do we hear a lot of sympathy for the investors that were invested when this bubble burst. No!

So do we need to be highly sympathetic (except on a humanitarian basis) for the people injured by Enron, WorldCom, etc.?

It was/is unfortunate, but they are not the only people injured when a bubble bursts. What about all of the people injured by the housing bubble. What about the people who had their pensions destroyed.

Bubbles come in a lot of shapes and sizes and they are difficult to spot in real time but they are very injurious to your financial health. You need to look at different types of bubbles. You will see that they do not always have a set of correlations that would allow you to spot them, other than the fact that at some point the underlying premise becomes "folk lore" and everyone is talking about this as a "truth" that will continue forever.

Chapter 34

DEFINITION OF BUBBLES

A bubble is a simple definition. A bubble is the inflating of asset values beyond what would be normal under most circumstances.

An easy way to get a feel for a bubble is to look at how fast the underlying assets are appreciating.

Assets can appreciate rapidly for short periods of time, but when assets appreciate for long periods of time, at very rapid rates, there will be trouble.

Apple recently created a bubble. It became the wealthiest company in the world. But it exists in a highly competitive market subject to normal revisions of its valuation (which happened during 2013). IBM during the late 80s and early 90s had the same characteristics. Tulips, although they had less intrinsic value, had a similar situation from 1636-1637. It was considered to be the first recorded financial bubble. Speculators pushed Tulip bulb prices to record heights before they collapsed and plunged the Dutch economy into a severe crisis that lasted for many years.

You can see lots of examples, but the examples all tend to run into an overvaluation based on some kind of mania. It's up to you to sort this out and avoid the mania and a bubble.

A key part of a bubble is rapid growth. Rapid growth by itself will not cause a bubble, but it will if that rapid growth is focused on a particular business segment. If the segment is easy to enter, you will get huge money flows and you will get a bubble. How the bubble will sort itself and the competitors out then becomes the only issue.

Unconstrained growth is just not possible and something will have to give. Yet, since you do not know the size of the market, you do not know when a bubble could break. You do not know how far asset values can appreciate. If you or others knew the answer then you would have spotted and acted upon the housing bubble. But, very few people acted even if they understood the problem.

This is true of most bubbles. You will not really know that it is a bubble until it has burst. Yet you can see the bubble puffing up. You may not be able to figure out if the bubble is close to bursting as you do not know how big it can get until after it bursts.

When assets begin to appreciate at a rapid rate, you need to beware of a bubble. If assets appreciate rapidly and people begin to think this is a natural and repeatable phenomenon then you need to be really concerned. If the appreciation starts to take on a long time line, then trouble is on the horizon.

At some point the, the assets must return to some level of normalcy.

Bubbles are a simple and inflated view of the worth of an asset caused by the desire of a lot of people to have and possess the same item. Even Carl Icahn, considered by many to be America's premier money guy, thinks the current stock market (late 2013) is overvalued and heading for a fall as its been propped up by easy money. When asked when the market is going down he was basically quoted as saying " I have no idea, nobody can tell you that." Even the smart guys can get caught in a bubble.[4]

---

4. Rana Foroohar. *The Original Wolf of Wall Stree*t. *Time*. 16 December 2013

CHAPTER 35
VISUALIZATION OF BUBBLES

Trying to visualize a bubble is not difficult. Just think of a simple soap bubble that you can blow from a pipe. This gives you the basic idea. The bubble is a fragile piece of material that will disappear at the slightest touch or will collapse if you try to blow it up too big. This is a good model to consider intellectually, but does not give you a basis to evaluate a product bubble.

From a product point of view, it would be nice if we could find a simple visualization of the formation of a bubble. It would be wonderful if this mechanism could give us an early hint that trouble may be in the future.

You never know how big a bubble will get before it pops. So you can never figure out the end point until after the fact, but you would like a hint that things are getting out of control.

A friend of mine (who is a stock trader) knows when things are getting out of hand. When he sees a bubble forming, he's got a saying, "It has gone exponential!"

That is what I look for. The surest sign that a bubble is forming is when the trend/wave/window starts to go exponential. If we are looking at a stock it looks like the price seems to levitate regardless of the circumstance and the stock price increases at an exponential rate. The same can be said of asset prices.

Suddenly, the price of gasoline jumps a dollar overnight and continues to rise. All of the rapid increases eventually become unsustainable. And, the market will correct the process and prick the bubble.

In the early 80s, I wrote a series of newsletters on the issue of Local Area Networks. There were literally hundreds of startup companies all claiming they would be successful if they could just get 1% of the market. I knew we were in trouble. Eventually there had to be a winner, but everyone was pouring in money on the basis of getting that 1%.

The internet bubble began when people believed that the cute dog puppet could deliver dog food with no shipping charges (in big, heavy and bulky bags). That's when you know that portions of the Internet commerce boom were about to collapse.

In the internet boom in the late 90s, 2000 time frame, people began to believe that you did not need to make a profit. You could just grow a commodity-based business, deliver goods and eventually you would be one of the lone survivors in that product space. No need to worry about making money along the way. At that point you knew the end had started.

It is difficult to spot the exact end of a bubble, but you can see warning signs—irrational exuberance anyone? While irrationality may be difficult to measure, the more people believe it does not exist, the further along the exponential growth curve we are and the shorter the time we have before the bubble will burst and some people will get their comeuppance.

It is not necessary to visualize the end. There is probably no way to actually measure and predict the end. But, you know it is coming. You can spot the signs of the demise. If you're tuned in, you can almost "feel" it. So you need to prepare for the end of the bubble and take steps to protect yourself.

If you think back to poking and popping the soap bubble, it may not happen as fast or be as dramatic and it may take a little longer, but popping a product or asset bubble can still cause a bang. The effect is the same. The product or asset essentially vanishes. Poof!

The trigger for this may be something simple. I once saw a stock go exponential and then it crashed. The stock crashed because it was way overvalued. But the measurement of that overvaluation was simple. As it traded higher, it ranged from 5 to 9 dollars a share. But the day it was above 10 dollars a share was the rapid start of its end. Think of the effect. Everyone knows the stock is overvalued. But, suddenly, instead of a stock whose value was in the single digit range, the stock suddenly became a stock whose value was in the two digit range. Poof! Everyone took another look and headed for the door at the same time.

A nice soap bubble came out of the pipe, went exponential, and broke.

Visualize what happens when a soap bubble pops; watch as it expands exponentially—because it will pop soon.

A classic example of a bubble that is familiar to most people is the housing market. **Figure 9** illustrates the number of new housing starts in the US for a recent period of time. It is interesting to see how fast the bubble can pop as the number of new starts declined almost 50% in a little over two years. But, at the same time the prices of existing houses also declined significantly—in many cases by 50% as well. When a bubble bursts there is no real way for you to get out of the way if you are invested in the bubble!

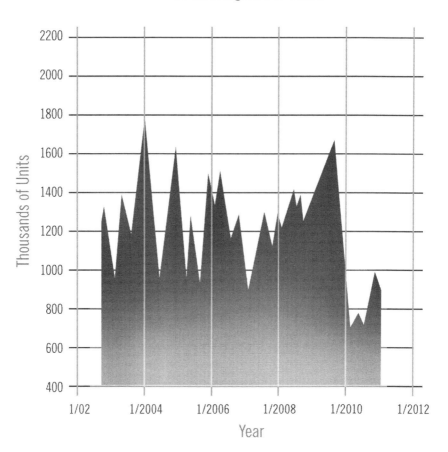

Figure 9 Illustration of Housing starts in the US from 2003 through 2012 (Source: Forecastchart.com, US Housing Starts)

Chapter 36
PROBLEMS WITH BUBBLES

The problem with bubbles was discussed in the previous chapter. They will pop. But, you can avoid being injured if you can evolve over time. If you assume that a product, technology or sales of a product will continue at a constant rate of growth you are setting yourself up for failure. Change is always occurring and you need to evolve constantly to avoid being caught as a bubble pops. You need to constantly be setting your future position in place.

Because of the amounts of money chasing outsized returns, anytime you have a developing industry and a potentially large profit you will attract lots of attention and you will face a bubble. It is guaranteed.

Many people view this as a problem. I do not see it as a problem. It is an issue of managing risk. Everything that you do involves risk. When you walk across a street, you are assuming risk. If you get in the shower you are assuming risk.

You need to understand that a bubble can ruin you. You need to treat it as a problem of risk management.

We live with risk every day and for common risks we have developed coping strategies. You are not bothered by the idea of walking across the street. You look left and right and wait for the red light and then you cross the street safely. You can do the same thing with bubbles.

You should not avoid a bubble. You should seek to protect yourself if you spot one. You need to develop alternative strategies. You can never tell how long a bubble will run or far it will go. You need to try to position yourself so that you need not fear the identified bubble. If you have and escape route or strategy you can try to ride the bubble as long as possible.

If you spot a bubble there are techniques you can use to protect yourself. If you own a stock that has suddenly gone exponential, you can cut back on your position. Or, you can set stop loss points. Or, you can buy puts and calls. There are a lot of techniques. The key is that when the bubble bursts, you can be seriously injured. So once you see (or even suspect) a bubble is forming you need to try to insulate yourself from the popping of the bubble.

There may be bubbles that do not pop. There may be ways to take the exuberance from an irrational position to a rational position, but I have not ever seen things end this way.

You need to understand that no matter how long and no matter how many people say it is different this time, it is not different this time. The bubble will pop. If/when you understand that, you can protect yourself.

Because every case is different, there is no sure way to protect yourself when the bubble pops. But, once you recognize that the bubble is growing exponentially, you can protect yourself.

Bubbles are funny. Periodically in the technology job market, someone decides that it is important that everyone who performs a certain job needs certification. At this point, people who have the certification can command a very rich salary premium. Once salaries start climbing due to this artificially induced shortage, lots of people pursue the certification and the price goes down. In that case, if you're lucky enough to have certification, you can do pretty well. However, eventually salaries will come back to earth. Save a little money. Do not overextend yourself. It is after all a simple business. Do not get caught in the formation of the bubble. Realize that the bubble will burst at some future point.

PROBLEMS WITH BUBBLES

Once you adopt this strategy, you have taken on another risk. When something is going exponential you do not know how far it will go. You should never assume that you will always be the first one through the window and last to profit before the window closes. You need to set reasonable expectations, but you need to understand that the growth will probably turn in directions you have not considered and further than you think. You need to develop strategies that will allow you to stay as long as possible and to profit as much as possible, but your strategy has to have elements that protect you when the bubble bursts.

The problem with bubbles is that you are really assuming two levels of risk. One has to do with your position when the bubble bursts and the other has to do with how much you left on the table. Both will be a problem and you need to figure out how to recognize and remedy both risk problems. Further complicating the problem is that the bubble may be changing as you as move through this process.

Chapter 37
Decay of Bubbles

Trends, waves and windows have different forms of decay that can be described. Bubbles on the other hand tend to pop. They end. That is not to say that a bubble gives you no warning when it pops. It may give you plenty of warning, but the result of a bubble is more like everyone trying to get through a door at the same time (with urgency). As people find it difficult to get through the door, they rush to get out at any cost. This is what causes the bubble to burst instead of decaying.

Whereas a trend may start to reverse, a bubble will tend to slow down. One of the fundamentals behind the formation of a bubble is that the bubble gets overdone and stretched to the limit so something has to give.

This does not mean that the bubble's growth will not slow down, but the process of a simple reversal of the trend or a simple change in direction in the rate of increase is not going to happen. The bubble has become conventional wisdom. Eventually, all of the money that is available to support the bubble has come in and there are no more people to support the bubble. Too many people recognize this at the same time and start to head for the exits. When everyone recognizes the situation, there is no way to get out as the bubble bursts.

That is not to say that it deflates instantly, but in comparison to the time it took to build the bubble, it does not have a long decay cycle and the bubble will implode quickly.

As an example, consider the housing bubble. The market got puffed up with liar loans and a number of people (and the people advising them) believed that a person with little or no income could live in a large house on essentially very little income. The theory was that houses would appreciate at a certain rate in the future. And, this rate was well above the historical rate. At the same time some genius figured out that you could decrease risk if you packaged a set of high risk loans together. (In retrospect, this was not a good mathematical model or assumption.) If you have a number of people who are all getting houses on the "cheap" because of future appreciation, what happens if the appreciation does not happen? If this drives up home ownership to unprecedented levels, you have the making of a disaster. We can call that 2007-2009. And, we are still sorting out the mess with massive numbers of foreclosures and the government bailout/bankruptcy of Fannie Mae and Freddie Mac.

On the surface it is not unreasonable to assume that house prices will go up. After all they do fluctuate. But it is not necessarily a good assumption that they will appreciate indefinitely at historic rates.

What you saw in the housing market was the housing market going exponential. The problem was you couldn't figure out when the game would stop. If it continued for a long time you want to be in the market. But, you wanted to be able to get out when it stopped. Unfortunately, that was difficult to time and lots of people got burned.

Suddenly everyone was getting a new house and the housing market showed rapid expansion for a number of years. The rate of new housing development began to slow as demand dried up. Finally, the bubble burst and prices dropped precipitously. In some hot markets, like Phoenix, Las Vegas and Florida, prices went down by over 50% in a period of a couple of years and it will take years (if ever) for prices to fully rebound.

DECAY OF BUBBLES

Lots of people ended up with houses under water (they owed more than the house was worth) and mortgage payments that were not sustainable.

However, if you look at the curve of house prices you will see gradual appreciation. Then the appreciation accelerated and suddenly you had the deflation of housing prices. The only hint that you had that the bubble was about to decay or burst is when house prices began to slow in their appreciation. Then the bubble burst quickly! And, the housing market seemed like it decayed as the market just continued to decline as if there was no end in sight. Yet if you were trying to sell a house you knew that the bubble had popped because there were just no buyers.

Chapter 38
DEATH OF A BUBBLE

A bubble popping is the death of the bubble. For some reason, it usually takes a reasonably long time for a bubble to get going. Then the bubble can take a long time to grow and fester. But when the bubble pops, it is a very quick process. And, recovery from a bubble popping is a long and tortuous process.

In most cases, when a bubble pops, it can take a lot of years to recover if in fact recovery is even possible. The problem with a bubble popping is that it leaves a bad taste in the mouths of a large number of people. This problem plus the overhang of product/assets that are now devalued makes recovery very difficult.

Think of it like a stock. If a stock gets kited and it goes exponential, there are lots of people hopping on the bandwagon to make that happen. If the stock then goes exponential and crashes from a massive run-up, there are a large number of people left holding the bag. If the stock ever tries to go up, there is an incredible overhang of stock held by people who are just trying to get even. They do not want to take a loss as that finalizes that they are victims and have lost money. It's human nature to continue to hold the stock and see if you can get even. This leaves an incredible overhang of stock just waiting to be sold if the stock gets above the level that the person bought it at. If the person gets even they think they never lost money. When in fact at some point they were in a losing position and by holding on they also are losing opportunities. But, you can feel good as until you sell the stock you have not "lost money".

Not recognizing that you have lost money in the market is one thing. However, if you were in the housing market, you were like a lot of people holding onto houses that were under water. And, this is really in your face. You have difficulty refinancing because the house is under water. At the same time you do not want to put more money into the house because you could use the money for something else. And, if you could get out from under the house you could live cheaper. In some cases, people with liar loans assumed they could afford the house because it would appreciate in value. This makes for a market with no place to go except sideways and that's when the house gets abandoned. This is a crisis as abandoned houses pull down neighborhoods and create potential for crime in the neighborhood. The problem gets compounded because it usually takes quite a while to foreclose on a house and you have people living in houses in which they have no vested interest. This will just tend to keep the pressure on the housing market for an extended period of time.

There is an old saying in the stock market that you should never try to catch a falling knife. And this is true when a bubble bursts. You have no idea about the size of the bubble or the financial holdings of the other people caught up in the bubble so you have no way of judging what will happen. You have no idea how far or fast the knife will fall. So everyone's tendency is to step aside and see how it works out. This will only tend to increase the size and speed of the fall. Eventually the asset becomes so underpriced that some number of people will step back into the asset class but because of the overhang, recovery is a long way off.

In the case of the stock market, a recent case is the stock of J.C. Penney where several large hedge funds have held strong positions and as the stock declined some groups shorted the stock and other groups took long positions and we had the case of greed and fear standing nose to nose and eye to eye while trying to beat each other to death.

You have a similar case in the housing market where big pools of money are stepping into the market and buying large portfolios of foreclosed houses to put into the rental market. The idea is that at some point the market will recover and the rental houses will be sold for a long term capital gain. In the meantime the houses will carry themselves through rental income.

Chapter 39
Variations of a Bubble

All bubbles are not the same. Once they pop, things can get really interesting. Some bubbles become susceptible to modification, some to just continuing on a downward spiral and some can provide new opportunities. Bubbles that can provide new opportunities are discussed in the next chapter on reinvention.

Bubbles that can be modified are very interesting. In many cases, the reason for the bubble leaves capabilities that can be exploited if you can figure out a cost strategy that will allow you to scrape out a living.

Consider the market for music. Music has been sold on vinyl, cassette tape, eight tracks, CDs and in digital files. Each of these delivery vehicles for the music has very different sound characteristics. Some people may not agree with the positions and assessments below, but these observations are based upon my experience.

Vinyl has the hiss and pop associated with a mechanical system and in many cases you cannot get rid of these side effects; but because the players for this technology are analog, you can get players that use tube technology to play the record, so you get a rich, warm sounding playback. That same warm effect is difficult, if not impossible, to get with a digital system. When you move up to cassette tape, you get a similar effect, but the breadth of the recording does not have as much overall frequency response due to limitations of the tape units. If you have an old fashion reel to reel system available, you might be able to approximate the vinyl sound. Additionally, with a cassette system, you have the effect of the Dolby noise reduction and it will have some effect on the total sound that you hear. As you move up to CDs you are in a place where the CDs could be analog or compressed digital and the sound can vary quite a bit. Moving all the way to compressed audio files, you will lose some of the sound and probably because the playback is digital you will lose some of the warmth of the sound.

ITunes and the advent of digital systems essentially destroyed the music market for everything but compressed digital music. But, because there is some number of people with a really good ear for hearing music, they want a higher sound quality. A small market has developed of people who buy and sell used equipment and content that is associated with the various playback and content systems listed above. This is just a situation where someone can position themselves in essentially a dead market and carve out the remainder of a very small niche for a period of time. This is a classic case of a market that has essentially recycled products and not much new product is actually developed. There is a small market in new vinyl products but it is very limited. Think of a typewriter repairman (they actually exist).

Other markets just continue to decay and eventually go out of business entirely as the technology becomes obsolete. An example of such a market is the market for a four function calculator. This market boomed in the mid-seventies. As calculators took off both in terms of price cutting and functions, the market for the simple calculator went to near zero. Today such calculators are usually just seen as advertising giveaways. The functions that were in the calculator now are just free functions in smart phones and computers. This is an example that for all practical purposes represents a market that just continued its downward spiral and ended up with no serious products and no value from a sales perspective. APPs on smart phones and tablets will probably do the same to even advanced calculators which will end up as nothing more than a simple piece of software running on a simple multi-purpose device.

As people begin to understand functions that can be programmed, as people understand how to build simple user interfaces and as hardware continues to advance, products that are in high demand tend to start into a downward sales cycle.

Chapter 40
REINVENTION OF BUBBLES

Not all bubbles end up being a disaster. In the previous chapter I said we would talk about bubbles that can provide new opportunities.

There are two ways this can happen. If the bubble that you are involved in has a lot of product that people want, the bubble really becomes a bubble due to pricing pressures. If we can become the low cost supplier we are in position to become a successful high volume producer of the product as long as we can keep coming out with new models and innovations. This is difficult and we must compete with new technologies that are trying to undercut our capabilities. But, we have the issue of inertia going for us. People generally do not like change and you will be surprised at how long we can milk a product line before we are out of business.

Another interesting thing to study is whether we can figure out how to reinvent a bubble.

In this case we are in the churn and probably the only way out is through another product opportunity. In that case we want to see if we can design some add-on features to high volume products being built by another manufacturer. This can become an interesting opportunity. This is called looking at the low side of the technology business.

The manufacturers who are trying to end up as the dominant

supplier in a commodity market must focus their attention on the base market and they must run fast! This means that they will not be looking at simple things like accessories or special high end functions that are of interest primarily to high end consumers.

High-end watch makers are a perfect example of this strategy. The function of a watch has been around for years. Now clocks are included in almost every device. Yet some watch makers still make mechanical devices that keep time and sell for serious prices. But these watches are not just a time piece they are jewelry. And, they contain packages that are outrageous in their extreme opulence. Just consider the watches produced by Patek Philippe. These watches can go for a lot of money and yet their basic function is still just—keeping time. The company distinguishes itself by emphasizing watches that "will delight connoisseurs and admirers of fine watch making" with "sublime artisanship".

Clearly, there is a market of high end consumers. And, for them a watch is more than just a time piece. It establishes position and status. Thus, Patek can survive in a market where there are tons of low cost alternatives. And, Patek is not the only provider of such products. There is a strong and vibrant market for such products. Yet you can go into to any discount store and find a wide choice of watches. There are even low to mid-range producers of such products that appeal to mass consumer audiences—Fossil watches and Swatch watches.

Just because a bubble has broken does not mean that all hope is lost. It just means that you must be very ingenious in figuring out how to position yourself.

One key is to find markets that are not dying but are moving into higher volume production and find a way to reinvent around the basic product.

Concluding Remarks on Bubbles

I find bubbles to be very interesting phenomena. They are the end result of people spotting a trend, a wave and a window and acting on their observation of the phenomena. But, there seems to be a herd mentality that magnifies the direction and duration of trends to the point that they eventually (and inevitably) get so puffed up that they will become economically unstable. That's when trouble will break lose.

This can be seen from tulip bulbs in 17th century Holland to the current housing crisis in the U.S. Part of the issue is that people do not understand compound arithmetic and eventually things become unsustainable. In other cases it is simply greed—using your home equity as an ATM—and bad assumptions about future asset growth, like home equity. In other cases, it is social engineering that may have allowed the housing crisis to go exponential when loan standards for mortgages were so relaxed that you did not have to prove you had income (or even if you actually had a job) to get a large mortgage.

Whatever the case, asset and product prices must revert to a normal valuation and the further the bubble has expanded the worse the resultant snap back or correction.

When substantial cash is pumped into the economy, asset bubbles can easily be formed.

But the real problem with bubbles formed by trends is that you never can be really sure how big the bubble is or how big the bubble will grow. No one wants to miss out on the profits so people eventually resort to the greater fool theory (there will always be someone willing to pay more than I did) and when that ceases to be true—POP—goes the bubble.

If you can get out in the early stages of deflation (or even as others are continuing to jump through the window) you can avoid the churn as the wave moves onto shore. Because people tend to be optimistic this is a skill that takes practice to learn.

EPILOGUE

In various ways over a long period of years, people have told me and I have observed that trends are important. However, there are more issues than simply looking at a trend. Consider the effects of waves, windows and bubbles—all of which we have been looking at in this book and all of which are byproducts of trends. A trend may last through a large number of waves and waves may provide many opportunities (or windows) to develop products and to profit. Unfortunately, a successful trend will most likely end in a bubble. It is human nature to expect good things will last forever.

A very recent example is the housing market collapse in the U.S. However, even in that disaster there is opportunity as now hedge funds and institutional investors are in the process of creating funds to buy, rehab and rent distressed and foreclosed houses. This is an interesting market because it may allow for the small individual investor to also profit if they are nimble. A carpenter friend of mine is actually taking this very approach—buy a foreclosed house, rehab it and rent it out.

Trends drive my thinking about products and sales strategies. There are many variations of advice about trends. A popular saying—"The trend is your friend"—is actually incomplete. One well-known and oft quoted version of the saying is "The trend is your friend, until it ends." You must recognize that trends can last for a long time, but they will eventually end or change the rate at which they are moving. In fact, in many cases the trend is actually a composite view of a set of observations which have been mathematically smoothed to illustrate their general direction. Nevertheless, trends provide us with guidance on the general direction of a specific area of interest and allow us to develop a strategy or framework for product development.

Trends that last a long time can provide multiple opportunities for developing and selling products. However, some trends are only viable for a short period of time. That is why you have to look at all the components, not just trends but waves, windows and bubbles.

Learning how to spot a wave that forms out of a trend is difficult, but that ability is invaluable for people who want to develop and sell products. The skill is invaluable for investors as well, particularly when trying to identify rapidly growing stocks before the competition discovers the growth. But, catching the wave (spotting the window) at the right spot to maximize profit is a very difficult skill to master. If you can figure out how to spot waves and exploit the window of opportunity you will be far ahead of the game.

Over the long term—and the long term seems to be getting shorter all of the time—trends, waves and windows will end badly. Bubbles are the result of success. It is human nature to believe that things are going to continue along their current path or trend. Couple that with the human trait to resist change and you have a situation where we do not recognize bubbles until they burst and usually that ends badly. People are optimistic and resistant to change and bubbles can be very harmful. Even people who can identify bubbles find it difficult to profit from their knowledge. You can never really judge how big a bubble will get before it bursts.

Your ability to spot a trend, wave or window is an important and valuable skill. It is the fundamental skill that will allow you to participate in the creation of products and wealth. However, just as important is your ability to control risk and avoid being run over by

EPILOGUE

a bubble that is out of control. In fact, one way to spot a real opportunity is to look for bubbles and to invest or go after bubble technologies—as long as you can control your risk exposure!

Since we do not know how big a bubble will get, there is no reason not to participate in a bubble as long as you think that your risks can be controlled. Risk control is a very difficult proposition and you need to be very careful.

Every day new trends are forming and waves are developing. If you can spot a window of opportunity and have the ability you should jump through the window and see what happens. There are many ways that you can participate as an investor or product developer, but you are doing yourself a disservice if you do not try to participate in a developing trend in some way or form.

Complex technology change provides opportunity. It changes the space of employment and the types and form of jobs. You must adapt or you will eventually be left behind. However, at any given time the rate of change may be small but adding up a lot of small changes eventually generates large opportunities.

Trends are forming as we speak and trends are causing waves. Your challenge is to find that window of opportunity and prosper. Whether it is a reversal of a trend (the housing implosion or the emergence of microbreweries) there is always opportunity available if you are looking for it. Seize the opportunities that are all around you because time is constantly moving.

Our challenge, as product developers or investors, is to spot trends and their resultant waves, jump through the window of opportunity and control our risks so we do not get crushed when the inevitable bubble bursts.

I hope this book has provided insight into the thinking of people who develop products and I hope you can prosper from that insight.

INDEX

3M, 62

Aiken, Howard, 25
Amazon, 58, 66
Analysts International, 26
Apple, 104
AppleWorks, 8
Apps, 91, 115

Bing, 108
Blackberry, 92, 118
Boston Beer Company, 73
Bricklin, Dan, 7
Bring your own device, 93
Bubbles
    Concept, 125
    Death, 147
    Decay, 143
    Definition, 133
    Importance, 127
    Problems, 139
    Reinvention, 155
    Types, 131
    Variations, 151
    Visualization, 135
BUNCH, 26
Burroughs, 26

CPT, 29
CSC, 26
Candy Crush, 50
cisco, 104
Control Data, 26

Darwin, Charles, 25
Dick Tracy, 15

Enron, 132

Facebook, 17, 34, 75, 122
Fad, 21
Fannie Mae, 144
Flint, 105
Fossil, 156
Frankston, Bob, 7
Freddie Mac, 144
Fry, Arthur, 62

GE, 83
GLF, 37
Gates, Bill, 62
Great Leap Forward, 37, 38
Google, 24

HD, 71, 126
Hamilton, Laird, 62
Honeywell, 26
Housing bubble, 125, 138

IBM, 12, 25
Icahn, Carl, 134
IPO, 122
iPad, 46
iPhone, 46, 91
iPod, 46

J. C. Penney, 148
Juniper Networks, 113

Kindle, 35, 93
Kiva, 58

Local area network, 86
Lotus 1-2-3, 8

Mainframe, 26
MAO, 37, 38
Madoff, Bernie, 132
Micro trend, 11
Microsoft, 8, 24, 104
Minicomputer, 26
Moore's Law, 2, 8, 20, 76
Motorola, 22
Multiplan, 8

NCR, 26
Netflix, 97

Office automation, 53
Oracle, 105
Osborne Computer Company, 43

PC, 27, 96
PDA, 22, 34
Patek Philippe, 156
Pet rock, 21
Post-it note, 62
Productivity, 57

Quality Record Pressings, 69
Quantum computing, 4

Rainbo Records, 69

SDC, 26
SSD, 9
Samsung, 15
Semi-log, 18
Smart phone, 22
Sony, 72
Software Arts, 7
Solid state drive, 9
Squire, 105
Stock market, 8
SuperCalc, 8
Swatch, 156
System 7, 12

INDEX

Tablets, 32

Tesla, 99

Traffic pattern, 3

Transistor density, 20

Trends
    Concept, 3
    Death, 29
    Decay, 25
    Definition, 13
    Importance, 7
    Problems, 21
    Reinvention, 37
    Types, 11
    Variations, 33
    Visualization, 17

Twitter, 75

UNIVAC, 26

VisiCalc, 7

Watson, Tom, 25

Waves
    Big, 46
    Concept, 45
    Death, 71
    Decay, 69
    Definition, 57
    Importance, 49
    Problems, 65
    Reinvention, 37
    Types, 53
    Variations, 75
    Visualization, 77

Windows
    Concept, 85
    Death, 111
    Decay, 107
    Definition, 95
    Importance, 89
    Problems, 103
    Reinvention, 117
    Types, 91
    Variations, 115
    Visualization, 99

WorldCom, 132

Made in the USA
San Bernardino, CA
17 June 2014